# Reliability Prediction from Burn-In Data Fit to Reliability Models

# Reliability Prediction from Burn-In Data Fit to Reliability Models

Joseph B. Bernstein
Ariel University

ELSEVIER

AMSTERDAM • BOSTON • HEIDELBERG • LONDON
NEW YORK • OXFORD • PARIS • SAN DIEGO
SAN FRANCISCO • SINGAPORE • SYDNEY • TOKYO
Academic Press is an imprint of Elsevier

Academic Press is an imprint of Elsevier
32 Jamestown Road, London NW1 7BY, UK
The Boulevard, Langford Lane, Kidlington, Oxford, OX5 1GB, UK
Radarweg 29, PO Box 211, 1000 AE Amsterdam, The Netherlands
225 Wyman Street, Waltham, MA 02451, USA
525 B Street, Suite 1900, San Diego, CA 92101-4495, USA

**Notices**
Knowledge and best practice in this field are constantly changing. As new research and
experience broaden our understanding, changes in research methods, professional practices, or
medical treatment may become necessary.

Practitioners and researchers must always rely on their own experience and knowledge in
evaluating and using any information, methods, compounds, or experiments described herein. In
using such information or methods they should be mindful of their own safety and the safety of
others, including parties for whom they have a professional responsibility.

To the fullest extent of the law, neither the Publisher nor the authors, contributors, or editors,
assume any liability for any injury and/or damage to persons or property as a matter of products
liability, negligence or otherwise, or from any use or operation of any methods, products,
instructions, or ideas contained in the material herein.

**British Library Cataloguing-in-Publication Data**
A catalogue record for this book is available from the British Library

**Library of Congress Cataloging-in-Publication Data**
A catalog record for this book is available from the Library of Congress

ISBN: 978-0-12-800747-1

For information on all Academic Press publications
visit our website at **store.elsevier.com**

This book has been manufactured using Print On Demand technology. Each copy is produced to
order and is limited to black ink. The online version of this book will show color figures where
appropriate.

Working together
to grow libraries in
developing countries

www.elsevier.com • www.bookaid.org

*For Rina,*
*my Ezer K'Negdi*

# CONTENTS

Introduction...........................................................................................ix

**Chapter 1 Shortcut to Accurate Reliability Prediction**...................1
1.1   Background of FIT.............................................................3
1.2   Multiple Failure Mechanism Model.....................................5
1.3   Acceleration Factor...........................................................6
1.4   New Proportionality Method...............................................9
1.5   Chip Designer..................................................................10
1.6   System Designer..............................................................13

**Chapter 2 M-HTOL Principles**..........................................................15
2.1   Constant Rate Assumption................................................15
2.2   Reliability Criteria...........................................................17
2.3   The Failure Rate Curve for Electronic Systems..................19
2.4   Reliability Testing...........................................................21
2.5   Accelerated Testing.........................................................25

**Chapter 3 Failure Mechanisms**..........................................................31
3.1   Time-Dependent Dielectric Breakdown..............................31
3.2   Hot Carrier Injection.......................................................38
3.3   Negative Bias Temperature Instability...............................42
3.4   Electromigration..............................................................44
3.5   Soft Errors due to Memory Alpha Particles.........................47

**Chapter 4 New M-HTOL Approach**....................................................49
4.1   Problematic Zero Failure Criteria.....................................50
4.2   Single versus Multiple Competing Mechanisms...................54
4.3   AF Calculation................................................................56
4.4   Electronic System CFR Approximation/Justification..............65
4.5   PoF-Based Circuits Reliability Prediction Methodology..........76
4.6   Cell Reliability Estimation................................................83
4.7   Chip Reliability Prediction...............................................87
4.8   Matrix Method.................................................................88

**Bibliography**......................................................................................93

# INTRODUCTION

This monograph represents over 15 years of research at the University of Maryland, College Park, in collaboration with Tel Aviv, Bar Ilan, and Ariel Universities in Israel. I am quite proud to present this unique approach for electronic system reliability assessment and qualification based on physics of failure combined with the collective knowledge of reliability professionals. This work illustrates a very straightforward and common sense approach to reliability assessment that includes the well-known property of constant rate failures that industry observes in the field with the mostly academic approach of physics-of-failure evaluation. Simply stated, we have developed a system by which a reliability engineer can take data from accelerated life testing which normally tests only a single mechanism by design and relates it to the proportion of constant rate failures that are observed in the field. Then, you can use the results of both field data and test data in order to more properly model the expected lifetime behavior of electronics as they operate under specified conditions.

This booklet is to be used by designers of highly complex electronic systems that are used in military, aerospace, medical, telecommunications, or any field where an accurate assessment of reliability is needed and can be verified through testing. I hope that we provide sufficient detail in our approach that can develop into a standard approach for reliability assessment and can be implemented at nearly any level of qualification from the device, component level up to the finished system. I want to give a great thank you to my students Moti Gabbay and Ofir Delly at Ariel University. I also want to express my greatest appreciation to Professor Shmuel Shternklar and the entire department of Electrical Engineering at Ariel University for allowing me to establish my laboratory there and for working to make this school a world recognized center for microelectronics reliability. I am also indebted to my former students from University of Maryland who worked with me on my first handbook, *Physics-of-Failure Based*

*Handbook of Microelectronic Systems*, including Shahrzad Salemi, Jin Qin, Xiaohu Li, and Liyu Yang.

The development of our approach described in the booklet has been funded by the Office of the Chief Scientist of Israel and the software that implements this method is available from BQR, at www.bqr.com.

# CHAPTER *1*

# Shortcut to Accurate Reliability Prediction

The traditional high-temperature operating life (HTOL) test is based on the outdated JEDEC standard that has not been supported or updated for many years. The major drawback of this method is that it is not based on a model that predicts failures in the field. Nonetheless, the electronics industry continues to provide data from tests of fewer than 100 parts, subjected to their maximum allowed voltages and temperatures for as many as 1000 h. The result based on zero, or a maximum of 1, failure out of the number of parts tested does not actually predict. This null result is then fit into an average acceleration factor (AF), which is the product of a thermal factor and a voltage factor. The result is a reported failure rate as described by the standard failure in time (FIT, also called Failure unIT) model, which is the number of expected failures per billion part hours of operation. FIT is still an important metric for failure rate in today's technology; however, it does not account for the fact that multiple failure mechanisms simply cannot be averaged for either thermal or voltage AFs.

One of the major limitations of advanced electronic systems qualification, including advanced microchips and components, is providing reliability specifications that match the variety of user applications. The standard HTOL qualification that is based on a single high-voltage and high-temperature burn-in does not reflect actual failure mechanisms that would lead to a failure in the field. Rather, the manufacturer is expected to meet the system's reliability criteria without any real knowledge of the possible failure causes or the relative importance of any individual mechanism. More than this, as a consequence of the nonlinear nature of individual mechanisms, it is impossible for the dominant mechanism at HTOL test to reflect the expected dominant mechanism at operating conditions, essentially sweeping the potential cause of failure under the rug while generating an overly optimistic picture for the actual reliability.

Two problems exist with the current HTOL approach, as recognized by JEDEC in publication JEP122G: (1) multiple failure mechanisms actually compete for dominance in our modern electronic devices

Reliability Prediction from Burn-In Data Fit to Reliability Models. DOI: http://dx.doi.org/10.1016/B978-0-12-800747-1.00001-1

and (2) each mechanism has a vastly different voltage and temperature AFs depending on the device operation. This more recent JEDEC publication recommends explicitly that multiple mechanisms should be addressed in a sum-of-failure-rates approach. We agree that a single point HTOL test with zero failures can, by no means, account for a multiplicity of competing mechanisms.

In order to address this fundamental limitation, we developed a special multiple-mechanism qualification approach that allows companies to tailor specifications to a variety of customers' needs. This approach will work with nearly any circuit to design a custom multiple HTOL (*M-HTOL*) test at multiple conditions and match the results with the foundrys' reliability models to make accurate FIT calculations based on specific customers' environments including voltage, temperature, and speed.

Fortunately, for today's sophisticated device manufacturer, we offer a unique and verifiable solution that gives the supplier a verifiably more accurate way to actually predict the expected field failure rate, FIT, based on the user's operating conditions. The chip foundry provides the manufacturer with very complex reliability calculators designed for each technology's process. The manufacturer then chooses specific accelerated tests that can be matched with the foundry's reliability models, and a simple solution for the expected failure rate (FIT) can be found for any user's expected applications. These models should be trusted by the manufacturer since they trust the models of the foundry. This way, there is confidence from the foundry to the user, and we can make believable reliability predictions in the field (Figure 1.1).

This approach fits the highly developed and sophisticated models of the foundry based on years of knowledge and testing of the physical failure mechanisms. These mechanisms are known inherently to lead to degradation and ultimately to a chip failure. The approach described herein puts this knowledge in the hands of the chip designers allowing for qualification at any expected operating conditions. This way, the same product may accommodate additional potential markets by selling the same design for higher reliability applications with broader operating margins or higher performance applications where long life is not as critical.

The traditional single-model HTOL gives an unrealistically low value for the expected FIT, and customers invariably find that their

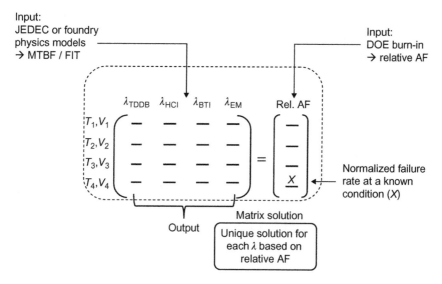

*Figure 1.1 Matrix methodology for reliability prediction.*

application shows a much higher reported failure rate than what was provided by the supplier using the traditional approach. Our M-HTOL matrix methodology will give a more accurate prediction for the expected field failure rate that is based on the actual test data and on the reliability models provided by the foundry to the chip designer. This way, the designer has a much more dependable picture for device reliability, and the customer will be satisfied that his design will match their customer's expectations for performance life.

## 1.1 BACKGROUND OF FIT

Reliability device simulators have become an integral part of the design process. These simulators successfully model the most significant physical failure mechanisms in modern electronic devices, such as time dependent dielectric breakdown (TDDB), negative bias temperature instability (NBTI), electromigration (EM), and hot carrier injection (HCI). These mechanisms are modeled throughout the circuit design process so that the system will operate for a minimum expected useful life.

Modern chips are composed of hundreds of millions or billions of transistors. Hence, chip-level reliability prediction methods are mostly statistical. Chip-level reliability prediction tools, today, model the failure probability of the chips at the end of life, when the known wearout mechanisms are expected to dominate. However, modern prediction tools do not predict the random, post burn-in failure rate that would be seen in the field.

Chip and packaged system reliability is still measured by a Failure-In-Time, alternatively called failure unIT (FIT). The FIT is a rate, defined as the number of expected device failures per billion part hours. An FIT is assigned for each component multiplied by the number of devices in a system for an approximation of the expected system reliability. The semiconductor industry provides an expected FIT for every product that is sold based on operation within the specified conditions of voltage, frequency, heat dissipation, etc. Hence, a system reliability model is a prediction of the expected mean time between failures (MTBFs) for an entire system as the sum of the FIT rates for every component.

An FIT is defined in terms of an AF as:

$$FIT = \frac{\#failures}{\#tested \times hours \times AF} \cdot 10^9 \tag{1.1}$$

where #failures and #tested are the number of actual failures that occurred as a fraction of the total number of units subjected to an accelerated test. The AF must be supplied by the manufacturer since only they know the failure mechanisms that are being accelerated in the HTOL, and it is generally based on a company proprietary variant of the MIL-HDBK-217 approach for accelerated life testing. The true task of reliability modeling, therefore, is to choose an appropriate value for AF based on the physics of the dominant failure mechanisms that would occur in the field for the device.

The HTOL qualification test is usually performed as the final qualification step of a semiconductor manufacturing process. The test consists of stressing some number of parts, usually around 77, for an extended time, usually 1000 h, at an accelerated voltage and temperature. Two features shed doubt on the accuracy of this procedure: one feature is lack of sufficient statistical data and the second is that companies generally present zero failure results for their qualification tests and hence stress their parts under relatively low stress levels to guarantee zero failures during qualification testing.

## 1.2 MULTIPLE FAILURE MECHANISM MODEL

Whereas the failure rate qualification has not improved over the years, the semiconductor industry understanding of reliability physics of semiconductor devices has advanced enormously. Every known failure mechanism is so well understood and the processes are so tightly controlled that electronic components are designed to perform with reasonable life and with *no single dominant failure mechanism*. Standard HTOL tests generally reveal multiple failure mechanisms during testing, which would suggest also that no single failure mechanism would dominate the FIT rate in the field. Therefore, in order to make a more accurate model for FIT, a preferable approximation should be that all failures are *equally likely*, and the resulting overall failure distribution resembles *constant failure rate process* that is consistent with the MIL handbook and FIT rate approach.

The acceleration of a single failure mechanism is a highly nonlinear function of temperature and/or voltage. The temperature acceleration factor ($AF_T$) and voltage acceleration factor ($AF_V$) can be calculated separately and is the subject of most studies of reliability physics. The total AF of the different stress combinations will be the product of the AFs of temperature and voltage,

$$AF = \frac{\lambda(T_2, V_2)}{\lambda(T_1, V_1)} = AF_T \cdot AF_V = \exp\left(\frac{E_a}{k}\left(\frac{1}{T_1} - \frac{1}{T_2}\right)\right)\exp(\gamma_1(V_2 - V_1))$$

(1.2)

This AF model is widely used as the industry standard for device qualification. However, it only approximates a single dielectric breakdown type of failure mechanism and does not correctly predict the acceleration of other mechanisms.

To be even approximately accurate, however, electronic devices should be considered to have several failure modes degrading simultaneously. Each mechanism "competes" with the others to cause an eventual failure. When more than one mechanism exists in a system, then the relative acceleration of each one must be defined and averaged at the applied condition. Every potential failure mechanism should be identified and its unique AF should then be calculated for each mechanism at a given temperature and voltage so the FIT rate can be approximated for each mechanism separately. Then the final FIT will be the sum of the failure rates per mechanism, as is described by:

$$FIT_{total} = FIT_1 + FIT_2 + \cdots + FIT_i \qquad (1.3)$$

where each mechanism leads to an expected failure unit per mechanism, $FIT_i$. Unfortunately again, individual failure mechanisms are not uniformly accelerated by a standard HTOL test, and the manufacturer is forced to model a single AF that cannot be combined with the known physics of failure models.

## 1.3 ACCELERATION FACTOR

Acceleration life testing of a product is often used to reduce test time by subjecting the product to a higher stress. The resulting data are analyzed and information about the performance of the product at normal usage condition is obtained. Acceleration testing usually involves subjecting test items to conditions more severe than encountered in normal use. This results in shorter test times, reduced costs, and decreased mean lifetimes for test times. In engineering applications, acceleration test conditions are produced by testing items at higher than normal temperature, voltage, pressure, load, etc. The data collected at higher stresses are used to extrapolate to some lower stress where testing is not feasible. In most engineering applications, a usage or design stress is known.

In the traditional approach, most models that can be used to derive the AF in actual use do not include any interaction terms, so that the relative change in AFs when only one stress changes does not depend on the level of the other stresses. The general Eyring model is almost the only model that includes terms that have stress and temperature interactions (in other words, the effect of changing temperature varies, depending on the levels of other stresses). Most models in actual use do not include any interaction terms, so that the relative change in AFs when only one stress changes does not depend on the level of the other stresses.

In models with no interaction, one can compute AFs for each stress and multiply them together. This would not be true if the physical mechanism required interaction terms. However, it could be used as the first approximations.

Taking the temperature and voltage acceleration effects independently, the overall AF can be obtained by multiplying the temperature acceleration factor by the voltage acceleration factor.

The traditional form for calculating AF is:

$$\text{AF}_{\text{sys}} = \prod_{i=1}^{k} \text{AF}_i, \quad \text{AF}_i = \text{acceleration factor each stress} \tag{1.4}$$

In general, the AF is provided by the manufacturers since only they know the failure mechanisms that are being accelerated in the HTOL, and it is generally based on a company proprietary variant of the MIL-HDBK-217 approach for acceleration life testing. The true task of reliability modeling, therefore, is to choose an appropriate value for AF based on the physics of the dominant device failure mechanisms that would occur in the field.

The qualification of device reliability, as reported by an FIT rate, must be based on an AF, which represents the failure model for the tested device. If we assume that there is no failure analysis (FA) of the devices after the HTOL test, or that the manufacturer will not report FA results to the customer, then a model should be made for the AF based on a combination of competing mechanisms. This will be explained by way of example. Suppose there are two identifiable, constant rate competing failure modes (assume an exponential distribution). One failure mode is accelerated only by temperature. We denote its failure rate as $\lambda_1(T)$. The other failure mode is only accelerated by voltage, and the corresponding failure rate is denoted as $\lambda_2(V)$.

By performing the acceleration tests for temperature and voltage separately, we can get the failure rates of both failure modes at their corresponding stress conditions. Then we can calculate the AF of the mechanisms. If for the first failure mode, we have $\lambda_1(T_1), \lambda_1(T_2)$ and for the second failure mode, we have $\lambda_2(V_1), \lambda_2(V_2)$, then the temperature acceleration factor is:

$$\text{AF}_T = \frac{\lambda_1(T_2)}{\lambda_1(T_1)}, \quad T_1 < T_2 \tag{1.5}$$

and the voltage acceleration factor is:

$$\text{AF}_V = \frac{\lambda_2(V_2)}{\lambda_2(V_1)}, \quad V_1 < V_2 \tag{1.6}$$

The system AF between the stress conditions of $(T_1, V_1)$ and $(T_2, V_2)$ is:

$$AF = \frac{\lambda_1(T_2, V_2) + \lambda_2(T_2, V_2)}{\lambda_1(T_1, V_1) + \lambda_2(T_1, V_1)} = \frac{\lambda_1(T_2) + \lambda_2(V_2)}{\lambda_1(T_1) + \lambda_2(V_1)} \quad (1.7)$$

The above equation can be transformed to the following two expressions:

$$AF = \frac{\lambda_1(T_2) + \lambda_2(V_2)}{\frac{\lambda_1(T_2)}{AF_T} + \frac{\lambda_2(V_2)}{AF_V}} \quad (1.8)$$

or

$$AF = \frac{\lambda_1(T_1)AF_T + \lambda_2(V_1)AF_V}{\lambda_1(T_1) + \lambda_2(V_1)} \quad (1.9)$$

These two equations can be simplified based on different assumptions.

When $\lambda_1(T_1) = \lambda_2(V_1)$ (i.e., equal probability at-use condition):

$$AF = \frac{AF_T + AF_V}{2} \quad (1.10)$$

Therefore, unless the temperature and voltage are carefully chosen so that $AF_T$ and $AF_V$ are very close, within a factor of about two, then one AF will overwhelm the failures at the accelerated conditions. Similarly, when $\lambda_1(T_2) = \lambda_2(V_2)$ (i.e., equal probability during accelerated test condition), then AF will take this form:

$$AF = \frac{2}{\frac{1}{AF_T} + \frac{1}{AF_V}} \quad (1.11)$$

and the AF applied to at-use conditions will be dominated by the individual factor with the smallest acceleration. In either situation, the accelerated test does not accurately reflect the correct proportion of AFs based on the understood physics of failure mechanisms.

This discussion can be generalized to incorporate situations with more than two failure modes. Suppose a device has $n$ independent failure mechanisms, $\lambda_{LTFMi}$ represents the $i$th failure mode at accelerated condition, and $\lambda_{useFMi}$ represents the $i$th failure mode at normal condition; then AF can be expressed. If the device is designed so that the failure modes have equal frequency of occurrence during the *use conditions*:

$$AF = \frac{\lambda_{useFM1} \cdot AF_1 + \lambda_{useFM2} \cdot AF_2 + \cdots + \lambda_{useFMn} \cdot AF_n}{\lambda_{useFM1} + \lambda_{useFM2} + \cdots + \lambda_{useFMn}} = \frac{\sum_{i=1}^{n} AF_i}{n}$$

$$(1.12)$$

If the device is designed so that the failure modes have equal frequency of occurrence during the *test conditions*:

$$AF = \frac{\lambda_{LTFM1} + \lambda_{LTFM2} + \cdots + \lambda_{LTFMn}}{\lambda_{LTFM1} \cdot AF_1^{-1} + \lambda_{LTFM2} \cdot AF_2^{-1} + \cdots + \lambda_{LTFMn} \cdot AF_n^{-1}} = \frac{n}{\sum_{i=1}^{n} \frac{1}{AF_i}}$$

(1.13)

From these relations, it is clear that only if AFs for each mode are almost equal, that is, $AF_1 \approx AF_2$, the total AF will be $AF = AF_1 = AF_2$ and certainly not the product of the two (as is currently the model used by industry). If, however, the acceleration of one failure mode is much greater than the second, the standard FIT calculation (Eq. (1.2)) could be incorrect by many orders of magnitude.

Due to the exponential nature of AF as a function of $V$ or $T$, if only a single parameter is changed, then it is not likely for more than one mechanism to be accelerated significantly compared to the others for any given $V$ and $T$. As we will see in Section 1.4, at least four mechanisms should be considered. In addition, the various voltage and temperature dependencies must be considered in order to make a reasonable reliability model for electron devices. Until now, the assumption of equal failure probability at-use conditions is used since it is the most conservative approach assuming the correct proportionality cannot be determined.

## 1.4 NEW PROPORTIONALITY METHOD

The basic method for considering multiple mechanisms as a system of simultaneous equations was first described in the paper from Bernstein et al. [1] and using the suggestion of a sum-of-failure-rate method as described in the JEDEC standard [2]. This same principle is discussed by Altera in their reliability report [3], although it is clear that they do not use this method for their reliability prediction. Similarly, the manufacturers of electronic components recognize the importance of combining failure mechanisms in a sum-of-failure-rates method. Also, the formulae for each mechanism is well studied and published in various places, including the IEEE International Reliability Physics Symposium. The prediction of microelectronic system reliability (matrix approach) is described.

The matrix approach is to model useful life failure rate (FIT) for components in electronic assemblies by assuming each component is composed of multiple subcomponents, for example, a certain percentage is effectively ring oscillator, static SRAM, and DRAM. Each type of circuit, based on its operation, can be seen to affect the potential steady-state (defect-related) failure mechanisms differently based on the accelerated environment, for example, EM, hot carrier, and NBTI. Each mechanism is known to have its own AFs with voltage, temperature, frequency, cycles, etc. Each subcomponent will be modeled to approximate the relative likelihood of each mechanism per subcomponent. Then, each component can be seen as a matrix of subcomponents, each with its own relative weight for each possible mechanism.

Hence, the standard system reliability FIT can be modeled using traditional MIL-handbook-217 type of algorithms and adapted to known system reliability tools; however, instead of treating each component as individuals, we propose treating each complex component as a series system of subcomponents, each with its own reliability matrix.

## 1.5 CHIP DESIGNER

A failure mechanism matrix can be solved at any given set of conditions, that is, voltage, temperature, and frequency, as percentages at each stressed operating condition; there is a unique proportion of each mechanism for a given set of stressed conditions that will result in the given time to fail (TTF). For example, the following matrix of calculated AFs are calculated relative to the unity condition, which is what we call the minimum use condition (MUC). In our example, this will be 0.9 V at 30°C operation.

In this matrix, the left-hand column describes the voltage temperature that is assumed for the accelerated testing condition (i.e., 1.6 V at 30°C). The relative AF based on the non-normalized formulae for each failure mechanism is given in each column that follows (i.e., HCI, BTI) (Table 1.1). These numbers assume equal contribution from each mechanism; however, the actual system will not have equal contributions for each mechanism. Instead, we need to test the product and extrapolate a relative TTF at each condition, and we want to compare in order to find a unique solution for percentages of each mechanism ($P_i$).

| Table 1.1 Relative AFs for Multiple Mechanisms Based on the Published Models from JEDEC or from the Foundry's Models Normalized to One Voltage and Temperature | | | | |
|---|---|---|---|---|
| Voltage_Temp | HCI | BTI | EM | TDDB |
| 1.6 V_30°C | 1,667,671 | 6,762,741 | 6 | 176,329,184 |
| 1.5 V_120°C | 275,664 | 279,314,765 | 523,745 | 14,658,341,454 |
| 1.3 V_120°C | 7532 | 5,115,828 | 523,745 | 200,284,634 |
| 1.4 V_120°C | 45,567 | 37,801,143 | 523,745 | 1,850,069,880 |
| 1.4 V_30°C | 45,567 | 123,864 | 6 | 3,210,449 |
| 0.9 V_30°C | 1 | 1 | 1 | 1 |

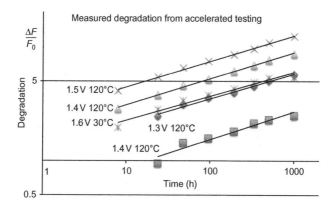

*Figure 1.2 Measured ring oscillator degradation at various test conditions.*

In order to find a relative failure rate for each mechanism, we take the accelerated life test at various voltages and temperatures, as shown in Figure 1.2.

For each condition, a consistent failure criterion must be chosen and the times to reach that condition yield "TTF" for the given conditions of voltage and temperature. The formula for percentage degradation is given as $y = \ldots$. Thus, solving each formula for the TTF (10% in our example) gives the following result.

In this example, the longest TTF is 211,917 h, so the relative AF is found by dividing the TTF for each condition by this value for 1.4 V at 30°C. The result is the middle row of Table 1.2 labeled AF. The final line is the expected FIT (failures per billion part hours) at those conditions.

Since the AF is relative to the still accelerated conditions, 1.4 V and 30°C (in our example), we must still find the relative acceleration compared to our minimal use conditions (MUCs) of 0.9 V at 30°C. In order to do this, we need to solve the system of equations from Table 1.2 by choosing three conditions from Table 1.2, multiplied by a new relative factor (RF) by which the longest TTF can be calibrated with MUC. The matrix is then solved for the first three conditions while substituting [1 1 1 1] for the MUC condition in the matrix.

This gives us a $4 \times 4$ matrix of accelerated conditions and a unity condition at 0.9 V and 30°C so we are sure that the sum of all the percentages ($P_i$) add up to unity (1) (Table 1.3).

There is a unique solution that gives us a result for the relative contributions (percentages) for each mechanism which will only be solved with a particular RF that assures that all the percentages are greater than 1 (Table 1.4). By imposing this condition, we found that the RF = 50,000 and is the relative voltage acceleration factor at 1.4 V from 0.9 V at 30°C.

| Table 1.2 Measured Times to 10% Degradation Based on Accelerated Test Data (Figure 1.2) | | | | | |
|---|---|---|---|---|---|
| Condition | 1.3 V 120°C | 1.4 V 30°C | 1.4 V 120°C | 1.5 V 120°C | 1.6 V 30°C |
| TTF | 12,046 | 211,917 | 1625 | 365 | 11,205 |
| AF | 18 | 1 | 130 | 580 | 19 |
| FIT | 83,012 | 4718 | 615,292 | 2,736,212 | 89,242 |

| Table 1.3 Relative AFs Compared to the Slowest Time to Fail (RF) | | |
|---|---|---|
| From Table 1.2 | Conditions | AF * RF |
| | 1.3 V_120°C | 18 RF |
| | 1.4 V_120°C | 130 RF |
| | 1.4 V_30°C | RF |
| | 0.9 V_30°C | 1 |

| Table 1.4 Results of the Matrix Solution Fitting Test Data with Models | | | |
|---|---|---|---|
| HCI | BTI | EM | GOI |
| 73.808118% | 9.038098% | 16.992687% | 0.161098% |

| Table 1.5 Normalized Test Data (Top Half) with Extrapolated Expected FIT (Right Column) | | | | | |
|---|---|---|---|---|---|
| | HCI | BTI | EM | GOI | FIT Final |
| 1.6_30 | 1,667,671 | 6,762,741 | 6 | 176,329,184 | 200,624.8 |
| 1.5_120 | 275,664 | 279,314,765 | 523,745 | 14,658,341,454 | 4,637,933.5 |
| 1.3_120 | 7532 | 5,115,828 | 523,745 | 200,284,634 | 82,997.7 |
| 1.4_120 | 45,567 | 37,801,143 | 523,745 | 1,850,069,880 | 615,185.5 |
| 1.4_30 | 45,567 | 123,864 | 6 | 3,210,449 | 4718.0 |
| 0.9 V_30°C | 1 | 1 | 1 | 1 | 0.1 |
| 0.9 V_70°C | 6 | 55 | 547 | 71 | 9.6 |
| 0.9 V_85°C | 6 | 131 | 3041 | 185 | 50.3 |
| 1.0 V_105°C | 34 | 3034 | 29,981 | 15,602 | 511.3 |
| 0.9 V_105°C | 6 | 411 | 29,981 | 661 | 484.7 |

By substituting these percentages into the matrix, the true AFs are determined for not only the tested condition but also for any extrapolated condition. The accelerated test formulae, which are not fixed by percentage contribution, are then multiplied by these percentages for any given actual use condition. Then, the relative FIT is fixed by the third row in Table 1.2, where 1.4 V at 30°C gives an actual FIT of 4718. At the same time, we saw that the AF for that same condition was 50,000 relative to 0.9 V at 30°C, which completes the matrix as follows (Table 1.5).

This completes our analysis for solving the reliability matrix in order to yield the FIT final (right-hand column) as a weighted sum of the mechanisms based on experimental verification using accelerated testing condition.

## 1.6 SYSTEM DESIGNER

A more sophisticated system that involves many chips, passives, and solder joints can use the same methodology as described earlier. The only complexity that is added by higher levels of failure mechanisms is the inclusion of more models. Of course, the testing becomes even more difficult since the degradation may not easily be detected. However, additional considerations, including packaging, solder joint integrity, moisture, and vibration, can all be added using current traditional reliability models. For example, solder joints are known to be dependent on thermal cycles more than on voltage or temperature, so this factor would be independent of the on-chip FIT. This model accounts for the failure rate of the chip as used in the packaging and in the system based on its operation, including voltage, temperature, and frequency, as it affects EM and hot carrier degradation.

In what follows, I will present the theoretical underpinnings as to how this method may be used and best applied to a system reliability evaluation by first justifying the constant rate, FIT approach for reliability evaluation. Then I will explain how the individual mechanisms are activated by the accelerated testing so we understand that it is impossible to activate only the most important mechanism that will lead to a failure with a single HTOL test at one accelerated voltage and temperature and that an M-HTOL is needed.

# CHAPTER 2

# M-HTOL Principles

Zero failures out of $N$ samples tested at a single accelerated voltage and temperature as the statistical foundation for building a reliability prediction by industry is one of the criteria that removes the believability of reliability predictions. One of the engineers' responsibilities is to do "conjecture" in order to find new ways to test not only the products but also the physical theories behind the tests. Instead, "competing failure mechanisms" is one of the suggestions, which could lose the "zero failure test" paradigm. Once we accept it (now that it has been accepted by more of the industry), we could apply the accelerating test matrix to provide accurate acceleration factors. Then, not only the accurate lifetime of system could be predicted, but also flaws and weaknesses could be revealed. In fact, this multiple HTOL prediction can be used as a sort of virtual failure analysis.

## 2.1 CONSTANT RATE ASSUMPTION

The most controversial aspects of the traditional approach could be classified as follows; the concept of constant failure rate assumption, the use of Arrhenius relation, the difficulty to maintain support data, problems of collecting good quality field data, and the diversity of failure rates of different mechanisms. These are especially true when the lack of provided data and other limitations presented with traditional approaches. Despite the disadvantages and limitations of traditional/ empirical-based handbooks, they are still used by engineers; strong factors such as good performance in the area of field reliability, ease of use, and providing an approximate field failure rate make them still popular. Crane survey shows that almost 80% of the respondents use MIL handbook, while PRISM and Telcordia have the second and third place [5].

The first publications on reliability predictions for electronic equipment were all based on curve fitting a mathematical model to historical field failure data in order to determine the constant failure rate of parts. None of them could include a root-cause analysis to the

Reliability Prediction from Burn-In Data Fit to Reliability Models. DOI: http://dx.doi.org/10.1016/B978-0-12-800747-1.00002-3

traditional/empirical approach. The physics-of-failure approach received a big boost when the US Army Material Command authorized a program to institute a transition from reliance exclusively on MIL-HDBL-217. The command's Army Material System Analysis Activity (Amsaa), Communications-Electronic Command (Cecom), the Laboratory Command (Lab-com), and the University of Maryland's CALCE group collaborated on developing a physics-of-failure hand-book for reliability assurance. The methodology behind the handbook was assessing system reliability on the basis of environmental and oper-ating stresses, the material used, and the packaging selected.

Two simulation tools, namely Computer-Aided Design of Microelectronic Packages (CADMP-2) and Computer-Aided Life-Cycle Engineering (CALCE), were developed to help the assessment; CADMP-2 assesses the reliability of electronics at the package level while CALCE assesses the reliability of electronics at the printed wiring board level. Together, these two models provide a framework to support a physics-of-failure approach to reliability in electronic systems design.

Physics of failure uses the knowledge of root-cause failure to design and to do the reliability assessment, testing, screening, and stress mar-gins in order to prevent product failures. The main task of a physics-of-failure approach is to identify potential failure mechanisms, failure sites, and failure modes and the appropriate failure models and their input parameters; determine the variability for each design parameter; and compute the effective reliability function. In summary, the objec-tive of any physics-of-failure analysis is to determine or predict when a specific end-of-life failure mechanism will occur for an individual com-ponent in a specific application.

A physics-of-failure prediction looks at each individual failure mech-anism, such as electromigration, solder joint cracking, and die bond adhesion, to estimate the probability of component wearout within the useful life of the product. This analysis requires detailed knowledge of all material characteristics, geometries, and environmental conditions. The subject is still modem, taking into account that new failure mechan-isms are discovered and even the old ones are not completely explained. One of the most important advantages of the physics-of-failure approach is the accurate prediction of wearout mechanisms. Moreover, since the acceleration test is one of the main aspects to find the model parameters, it could also provide the necessary test criteria for the product. To sum up, modeling of potential failure mechanisms, predicting the end of life, and using generic failure models effective for new materials and structures are the achievements of this approach [2].

The disadvantages of physics-of-failure approaches are related to their cost, the complexity of combining the knowledge of materials, process, and failure mechanisms together, the difficulty in estimating the field reliability, and their inapplicability to the devices already in production as the result of their incapability of assessing the whole system [2,6]. The method needs access to the product materials, process, and data.

Nowadays circuit designers have reliability simulations as an integral part of the design tools, like Cadence Ultrasim and Mentor Graphics Eldo. These simulators model the most significant physical failure mechanisms and help the designers to meet the lifetime performance requirement. However, there are disadvantages that hinder designers to adopt these tools. First, the tools are not fully integrated into the design software because the full integration requires technical supports from both the tool developers and the foundry. Second, they cannot handle the large-scale design efficiently. The increasing complexity makes it impossible to exercise full-scale simulation, considering the resources that simulation will consume. Chip-level reliability prediction only focuses on the chip's end of life, while the known wearout mechanisms are dominant; however, these prediction tools do not predict the random, post burn-in failure rate that would be seen in the field [7,8].

## 2.2 RELIABILITY CRITERIA

Reliability is defined as the probability that a system (component) will function over some time period $t$. If $T$ is defined as a continuous random variable, the time to failure of the system (component), then reliability can be expressed as:

$$R(t) = \Pr\{T \geq t\} \quad \text{where } R(t) \geq 0, \ R(0) = 1, \ \lim_{t \to \infty} R(t) = 0 \quad (2.1)$$

For a given value of $t$, $R(t)$ is the probability that the time to failure is greater than or equal to $t$. $F(t)$ could be defined as:

$$F(t) = 1 - R(T) = \Pr\{T < t\}, \ F(0) = 0, \ \lim_{t \to \infty} F(t) = 1 \quad (2.2)$$

Then $F(t)$ is the probability that a failure occurs before $t$.

$R(t)$ is referred as the probability function and $F(t)$ as the cumulative distribution function (CDF) of the failure distribution. A third function defined by:

$$f(t) = \frac{dF(t)}{dt} = -\frac{dR(t)}{dt} \tag{2.3}$$

is called the probability density function (PDF) and has two properties:

$$f(t) \geq 0, \quad \int_0^\infty f(t)dt = 1 \tag{2.4}$$

Both the reliability function and the CDF represent areas under the curve defined by $f(t)$. Therefore, since the area beneath the entire curve is equal to one, both the reliability and the failure probability will define so that:

$$0 \leq R(t) \leq 1, \quad 0 \leq F(t) \leq 1 \tag{2.5}$$

The mean time to failure (MTTF) is defined by:

$$\text{MTTF} = E(T) = \int_0^\infty tf(t)dt \tag{2.6}$$

which is the mean or expected value of the probability distribution defined by $f(t)$.

Failure rate or hazard rate function provides an instantaneous rate of failure:

$$\Pr\{t \leq T \leq t + \Delta t\} = R(t) - R(t + \Delta t) \tag{2.7}$$

The conditional probability of a failure in the time interval from $t$ to $t + \Delta t$ given that the system has survived to time $t$ is:

$$\Pr\{t \leq T \leq t + \Delta t \mid T \geq t\} = \frac{R(t) - R(t + \Delta t)}{R(t)} \tag{2.8}$$

and then:

$$\frac{R(t) - R(t + \Delta t)}{R(t)\Delta t} \tag{2.9}$$

is the probability of failure per unit of time (failure rate). Set:

$$\lambda(t) = \lim_{\Delta t \to \infty} \frac{-[R(t + \Delta t) - R(t)]}{\Delta t} \cdot \frac{1}{R(t)} = \frac{-dR(t)}{dt} \cdot \frac{1}{R(t)} = \frac{f(t)}{R(t)} \tag{2.10}$$

| Table 2.1 The Relationship Among the Measures [1] | | | | | |
|---|---|---|---|---|---|
| | F(t) | R(t) | f(t) | $\lambda$(t) | |
| $F(t) =$ | $F(t)$ | $1 - R(t)$ | $\int_0^t f(x)\mathrm{d}x$ | $1 - \exp\left(-\int_0^t \lambda(x)\mathrm{d}x\right)$ | Probability of failure |
| $R(t) =$ | $1 - F(t)$ | $R(t)$ | $\int_t^\infty f(x)\mathrm{d}x$ | $\exp\left(-\int_0^t \lambda(x)\mathrm{d}x\right)$ | Reliability |
| $f(t) =$ | $\frac{\mathrm{d}F(t)}{\mathrm{d}t}$ | $\frac{\mathrm{d}R(t)}{\mathrm{d}t}$ | $f(t)$ | $\lambda(t)\exp\left(-\int_0^t \lambda(x)\mathrm{d}x\right)$ | PDF |
| $\lambda(t) =$ | $\frac{\frac{\mathrm{d}F(t)}{\mathrm{d}t}}{1 - F(t)}$ | $\frac{\mathrm{d}(\ln R(t))}{\mathrm{d}t}$ | $\frac{f(t)}{\int_t^\infty f(x)\mathrm{d}x}$ | $\lambda(t)$ | Failure rate |

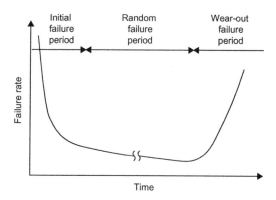

Figure 2.1 Failure rate curve.

Then $\lambda(t)$ is known as the instantaneous hazard rate or failure rate function. The failure rate function $\lambda(t)$ provides an alternative way of describing a failure distribution. Failure rates in some cases may be characterized as increasing (IFR), decreasing (DFR), or constant (CFR) when $\lambda(t)$ is an increasing, decreasing, or constant function. Table 2.1 gives the relationship among the measures.

## 2.3 THE FAILURE RATE CURVE FOR ELECTRONIC SYSTEMS

The reliability of electronic devices is represented by the failure rate curve (called the "bathtub curve"). The curve can be divided into the three following regions: (1) initial failures, which occur within a relatively short time after a device starts to be used, (2) random failures, which occur over a long period of time, and (3) wear-out failures, which increase as the device nears the end of its life (Figure 2.1).

"Initial failures" are considered to occur when a latent defect is formed, for example, during the device production process, and then become manifest under the stress of operation. For example, a defect can be formed by having tiny particles in a chip in the production process, resulting in a device failure later. The failure rate tends to decrease with time because only devices having latent defects will fail, and these devices are gradually removed. At the start of device operation, there may be electronic devices with latent defects included in the set. These will fail under operating stress, and as they fail, they are removed from the set based on the definition of the failure rate. The initial failure period can therefore be defined as the period during which devices having latent defects fail. The failure rate can be defined as a decreasing function, since the number of devices having latent defects decreases as they are removed.

Most of the microelectronic device initial defects are caused by defects built into devices mainly in the wafer process. The most common causes of these defects are dust adhering to wafers in the wafer process and crystal defects in the gate oxide film or the silicon substrate, etc. The defective devices surviving in the sorting process (during the manufacturing process) may be shipped as passing products. These types of devices, which are inherently defective from the start, often fail when stress (voltage, temperature, etc.) is applied for a relatively short period. Finally, almost all of these types of devices fail over a short period of time and the failure rate joins the region called random failure part. The process of applying stresses for a short period of time (before shipping the product) to eliminate the defective devices is called screening ("burn-in").

"Random failures" occur once devices having latent defects have already failed and been removed. In this period, the remaining high-quality devices operate stably. The failures that occur during this period can usually be attributed to randomly occurring excessive stress, such as power surges, and software errors. Moreover, memory software errors and other phenomena caused by $\alpha$-rays and other high-energy particles are sometimes classified as randomly occurring failure mechanisms. There are also phenomena such as Electro-Static Discharge (ESD), breakdown, overvoltage (surge) breakdown Electrical Over-Stress (EOS), and latch-up, which occur at random according to the conditions of use. However, these phenomena are all produced by the application of excessive stress over the device absolute maximum ratings, so these are classified as breakdowns instead of failures, and are not included in the random failure rate.

"Wear-out failures" occur due to the aging of devices from wear and fatigue. The failure rate tends to increase rapidly in this period. Semiconductor devices are therefore designed so that wear-out failures will not occur during their guaranteed lifetime. Accordingly, for the production of highly reliable electronic devices, it is important to reduce the initial failure rate to ensure the long life or durability against wear-out failures. Wear-out failures are failures rooted in the durability of the materials comprising semiconductor devices and the transistors, wiring, oxide films, and other elements, and are an index for determining the device life (useful years). In the wear-out failure region, the failure rate increases with time until ultimately all devices fail or suffer characteristic defects.

## 2.4 RELIABILITY TESTING

Reliability testing is a series of laboratory tests carried out under known stress conditions to evaluate the life span of a device or a system. Reliability tests are performed to ensure that semiconductor devices maintain performance and functions throughout their life. These reliability tests aim to simulate and accelerate the stresses that the semiconductor device may encounter during all phases of its life, including mounting, aging, field installation, and operation. The typical stress conditions are defined in the testing procedure described later in this document. Reliability tests are performed at various stages of design and manufacture. The purpose and contents differ in each stage. When testing semiconductor devices, (1) the subject and purpose of each test, (2) the test conditions, and (3) the judgment based on test results must be considered.

A number of standards have been established including Japan Electronics and Information Technology Industries Association (JEITA) standards, US Military (MIL) standards, International Electrotechnical Commission (IEC) standards, and Japan Industrial Standards (JIS). The testing procedures and conditions differ slightly from one another, but their purpose is the same. The test period including the number of cycles and repetitions and test conditions are determined in the same manner as the number of samples, by the quality level of design and the quality level required by the customer. However, mounting conditions, operating periods, and the acceleration of tests are considered to select the most effective and cost-efficient conditions. Reliability tests must be reproducible. It is preferable to select a standardized testing method. For this reason, tests are carried out according to international or national test standards.

The definition of a device failure is important in planning reliability tests. It is necessary to clarify the characteristics of the device being tested and set failure criteria. These failure criteria must be used in determining whether or not any variations in the device characteristics before and after the test are in an allowable range. The reliability of a device is usually determined by its design and manufacturing process. Therefore, a reliability test is conducted by taking samples from a population within the range that the factors of design and manufacturing process are the same. The sampling standard for semiconductor devices is determined by both the product's reliability goal and the customer's reliability requirements to achieve the required level of lot tolerance percent defective.

Reliability tests for materials and processes are performed with the number of samples determined independently. For newly developed processes and packages, however, there are cases in which the existing processes or packages cannot be accelerated within the maximum product rating or in which new failures cannot be detected within a short period of time. In these cases, it is important to perform reliability testing based on the failure mechanism by means of test element groups (TEGs). To ensure reliability during product design, we conduct testing according to the failure mechanism in the same process as the products, clarify the acceleration performance for temperature, electric field, and other factors, and reflect this in the design rules used in designing the product [9].

Table 2.2 lists the subject, purposes, and contents of some reliability tests carried out at some laboratories.

The standards and specifications related to the reliability of semiconductor devices can be classified as shown in [9,10]. Reliability methods are subject to standardization in several organizations. In those cases, the standards were usually derived from a common source (e.g., MIL-STD-883), and only minor differences exist between different standards (Table 2.3).

The MIL standards have been used as industry standards for many years, and today many companies still refer to the MIL standards. With

**Table 2.2 Examples of Reliability Testing Conducted When New Products Are Developed [1]**

| Phase | Purpose | Name | Contents |
|---|---|---|---|
| Development of semiconductor products | To verify that the design reliability goals and the customer's reliability requirements are satisfied | Quality approval for product developed | The following and other tests are carried out as required: 1. Standard tests 2. Accelerated tests 3. Marginal tests 4. Structural analysis |
| Development or change of materials and processes | To verify that materials and processes used for the product under development satisfy the design reliability goals and the customer's reliability requirements  To understand the quality factors and limits that are affected by materials and processes | Quality approval for wafer process/ package: development or change | TEGs or products are used to perform acceleration tests and other analyses as required with attention paid to the characteristics and changes of materials and processes |
| Pilot run before mass production | To verify that the production quality is at the specified level | Quality approval for mass production | This category covers reliability tests for examining the initial fluctuation of parameters that require special attention, as well as fluctuations and stability in the initial stage if mass production |

**Table 2.3 Classification of Test Standards**

**Classification of Standards**

| Class | Example |
|---|---|
| International standards | IEC |
| National standards | USA (ANSI) |
| Governmental standards | MIL |
| Industrial standards | EIA/JEDEC |
| Standards by expert groups | ESDA |
| Standards by semiconductor users | AEC Q100 |

the advent of new technologies, other standards have become more important. MIL-STD-883 establishes test methods and other procedures for integrated circuits. It provides general requirements such as mechanical, environmental, durability and electrical test methods. This document also provides quality assurance methods and procedures for procurement by the US military. The test methods described in MIL-STD-883 include environmental test methods, mechanical test methods, digital electrical test methods, linear electrical test methods, and test procedures. MIL-STD-750 specifies mechanical, environmental, and electrical test methods and conditions for discrete semiconductor devices. MIL-STD-202 specifies mechanical and environmental test methods for electronic and electrical components. MIL-HDBK-217 is a US Military handbook for reliability prediction of electronic equipment. This handbook, although long ago abandoned, introduces methods for calculating the predicted failure rate for each category of parts ion electronic equipment. Categories include mechanical, components, lamps, and advanced semiconductor devices, such as microprocessors.

The IEC was founded in 1908 and is the leading global organization that prepares and publishes international standards for electrical, electronic, and related technologies. IEC standards serve as a basis for national standardization and as references when drafting international tenders and contracts. IEC publishes a.o. international standards (IS), technical specifications (TS), technical reports (TR), and guides. Examples of reliability-related IEC standards are IEC 60068 (which deals with environmental testing; this standard contains fundamental information on environmental testing procedures and severities of tests and is primarily intended for electrotechnical products), IEC 60749 series (deals with mechanical and climatic test methods; these standards are applicable to semiconductor devices including discrete and integrated circuits), IEC 62047 (about microelectromechanical devices), IEC 62373/74 on wafer-level reliability, and IEC/TR 62380 about reliability data handbook universal model for reliability prediction of electronics components, printed circuit boards (PCBs), and equipment.

JEDEC Solid State Technology Association (formerly Joint Electron Device Engineering Council) is the semiconductor engineering standardization division of the Electronic Industries Alliance (EIA), a US-based trade association that represents all areas of the electronics industry. JEDEC was originally created in 1960 to cover the standardization of discrete semiconductor devices and later expanded in 1970 to include integrated circuits.

JEDEC spans a wide range of standards related to reliability. These standards are among the most referenced by semiconductor companies in the United States and Europe. The JEDEC committees for reliability standardization are JC-14 Committee on quality and reliability of solid-state products, JC-14.1 Subcommittee about reliability test methods for packaged devices, JC-14.2 Subcommittee on wafer-level reliability, JC-14.3 Subcommittee about silicon devices reliability qualification and monitoring, JC-14.4 Subcommittee on quality processes and methods, JC-14.6 Subcommittee about failure analysis, and JC-14.7 Subcommittee on gallium arsenide reliability and quality standards. JEDEC publishes standards, publications, guidelines, standard outlines, specifications, and ANSI and EIA standards.

JEITA was formed in 2000 from a merge of The Electronic Industries Association of Japan (EIAJ) and Japan Electronic Industry Development Association (JEIDA). Its objective is to promote the healthy manufacturing, international trade, and consumption of electronics products. JEITA's main activities are supporting new technological fields, developing international cooperation, and promoting standardization. Principal product areas covered by JEITA are digital home appliances, computers, industrial electronic equipment, electronic components, and electronic devices (e.g., discrete semiconductor devices and Integrated Circuits). JEITA standards related to semiconductor reliability are ED4701 which contains environmental and endurance test methods, ED4702 about mechanical tests, ED4704 on failure mode driven tests (WLR), EDR4704 about guidelines for acceleration testing, EDR4705 on SER, and EDR4706 about FLASH reliability; these standards are available online.

## 2.5 ACCELERATED TESTING

As mentioned by JEDEC, acceleration testing is a powerful tool that can be effectively used in two very different ways: in a qualitative or in a quantitative manner. Qualitative acceleration testing is used primarily to identify failures and failure modes while quantitative acceleration testing is used to make predictions about a product's life characteristics (e.g., MTTF, B10 life) under normal use conditions. In acceleration testing, the quantitative knowledge builds upon the qualitative knowledge. Using acceleration testing in a quantitative manner requires a physics-of-failure approach, that is, a comprehensive understanding and application of the specific failure mechanism involved and the relevant activating stress(es).

In general, acceleration lifetime tests are conducted under stress conditions more severe than actual conditions of use (fundamental conditions). They are methods of physically and chemically inducing the failure mechanisms to assess, in a short time, device lifetime and failure rates under actual conditions of use.

Increasing the degree of stressors (e.g., temperature, voltage), increasing the frequency of the applied stress, using tighter failure criteria, and using test vehicles/structures specific to the failure mode are the means of acceleration.

The stress methods applied in acceleration lifetime tests are constant stress and step stress. The constant stress method is a lifetime test where stress, such as temperature or voltage, is held constant and the degree of deterioration of properties and time-to-failure lifetime distribution are evaluated. In the step stress method, contrary to the constant stress method, the time is kept constant and the stress is increased in steps and the level of stress causing failure is observed.

Typical examples of tests using the constant stress method, the step stress method, and the cyclic stress method, a variation of the constant stress method, are given in Table 2.4.

Constructing an accurate quantitative acceleration test requires defining the anticipated failure mechanisms in terms of the materials used in the product to be tested, determining the environmental stresses to which the product will be exposed when operating and when not operating or stored. When choosing a test or combination of test on which a failure mechanisms based, assuming mechanisms that are anticipated to limit the life of the product, proper acceleration models must be considered. These generally include:

- Arrhenius Temperature Acceleration for temperature and chemical aging effects
- Inverse Power Law for any given stress
- Miner's Rule for linear accumulated fatigue damage
- Coffin–Manson nonlinear mechanical fatigue damage
- Peck's Model for temperature and humidity combined effects
- Eyring/Black/Kenney models for temperature and voltage acceleration

Table 2.5 describes each of these models, their relevant parameters, and frequent applications of each.

## Table 2.4 Typical Example of Acceleration Lifetime Tests

| Applied Stress Method | Purpose | Accelerated Test | Main Stressor | Failure Mechanism |
|---|---|---|---|---|
| Constant stress method | Investigation of the effects of constant stress on a device | High-temperature storage test | Temperature | Junction degradation, impurities deposit, ohmic contact, intermetallic chemical compounds |
| | | Operating lifetime test | Temperature | Surface contamination, junction degradation, mobile ions, Electromigration, etc. |
| | | | Voltage | |
| | | | Current | |
| | | High-temperature, high-humidity storage | Temperature | Corrosion, surface contamination, pinhole |
| | | | Humidity | |
| | | High-temperature, high-humidity bias | Temperature | Corrosion, surface contamination, junction degradation, mobile ions |
| | | | Humidity | |
| | | | Voltage | |
| Cyclic stress method | Investigation of the effects of repeated stress | Temperature cycle | Temperature difference | Cracks, thermal fatigue, broken wires and metallization |
| | | | Duty cycle | |
| | | Power cycle | Temperature difference | Insufficient adhesive strength of ohmic contact |
| | | | Duty cycle | |
| | | Temperature–humidity cycle | Temperature difference | Corrosion, pinhole, surface contamination |
| | | | Humidity difference | |
| Step stress method | Investigation of the stress limit that a device can withstand | Operation test | Temperature | Surface contamination, junction degradation, mobile ions, EMD |
| | | | Voltage | |
| | | | Current | |
| | | High-temperature reverse bias | Temperature | Surface contamination, junction degradation, mobile ions, TDDB |
| | | | Voltage | |

| Table 2.5 Frequently Used Acceleration Models, Their Parameters, and Applications | | | |
|---|---|---|---|
| **Model Name** | **Description/ Parameters** | **Application Examples** | **Model Equation** |
| Arrhenius Acceleration Model | Life as a function of temperature or chemical aging | Electrical insulation and dielectrics, solid state and semiconductors, intermetallic diffusion, battery cells, lubricants and greases, plastics, incandescent lamp filaments | $\text{Life} = A_0 e^{-\frac{E_a}{kT}}$ where: Life, median life of a population $A_0$, scale factor determined by experiment $e$, base of natural logarithms $E_a$, activation energy (unique for each failure mechanism) $k$, Boltzmann's constant $= 8.62 \times 10^{-5}$ eV/K $T$, temperature (degrees Kelvin) |
| Inverse Power Law | Life as a function of any given stress | Electrical insulation and dielectrics (voltage endurance), ball and roller bearings, incandescent lamp filaments, flash lamps | $\frac{\text{Life at normal stress}}{\text{Life at accelerated stress}} = \left(\frac{\text{Accelerated stress}}{\text{Normal stress}}\right)^N$ where: $N$, acceleration factor |
| Miner's Rule | Cumulative linear fatigue damage as a function of flexing | Metal fatigue (valid only up to the yield strength of the material) | $\text{CD} = \sum_{i=1}^{k} \frac{C_{Si}}{N_i} \leq 1$ where: CD, cumulative damage $Cs$, number of cycles applied at stress Si $N_i$, number of cycles to failure under stress Si (determined from an S−N diagram for that specific material) $k$, number of loads applied |
| Coffin−Manson | Fatigue life of metals (ductile materials) due to thermal cycling and/or thermal shock | Solder joints and other connections | $\text{Life} = \frac{A}{(\Delta T)^B}$ where: Life, cycles to failure $A$, scale factor determined by experiment $B$, scale factor determined by experiment $\Delta T$, temperature change |
| Peck's Model | Life as a combined function of temperature and humidity | Epoxy packaging | $\tau = A_0(\text{RH})^{-2.7}\exp\left[\frac{0.79}{kT}\right]$ where: $t$, median life (time to failure) $A_0$, scale factor determined by experiment RH, relative humidity $k$, Boltzmann's constant $= 8.62 \times 10^{-5}$ eV/K $T$, temperature (degrees Kelvin) |

*(Continued)*

## Table 2.5 (Continued)

| Model Name | Description/ Parameters | Application Examples | Model Equation |
|---|---|---|---|
| Peck's Power Law | Time to failure as a function of relative humidity, voltage, and temperature | Corrosion | $TF = A_0 \cdot RH^{-N} \cdot f(V) \cdot \exp[E_a/kT]$ <br><br> where: <br><br> TF, time to failure <br> $A_0$, scale factor determined by experiment <br> RH, relative humidity <br> $N = \sim 2.7$ <br> $E_a = 0.7\text{--}0.8$ eV (appropriate for aluminum corrosion when chlorides are present) <br> $f(V)$, an unknown function of applied voltage <br> $k$, Boltzmann's constant = $8.62 \times 10^{-5}$ eV/K <br> $T$, temperature (degrees Kelvin) |
| Eyring/Black/ Kenney | Life as a function of temperature and voltage (or current density (Black)) | Capacitors, electromigration in aluminum conductors | $\tau = \frac{A}{T} \exp\left[\frac{B}{kT}\right]$ <br><br> where: <br><br> $t$, median life (time to failure) <br> $A$, scale factor determined by experiment <br> $B$, scale factor determined by experiment <br> $k$, Boltzmann's constant = $8.62 \times 10^{-5}$ eV/K <br> $T$, temperature (degrees Kelvin) |
| Eyring | Time to failure as a function of current, electric field, and temperature | Hot carriers injection, surface inversion, mechanical stress | $TF = B(I_{sub})^{-N} \exp(E_a/kT)$ <br><br> where: <br><br> TF, time to failure <br> $B$, scale factor determined by experiment <br> $I_{sub}$, peak substrate current during stressing <br> $N = 2\text{--}4$ <br> $E_a = -0.1$ to $-0.2$ eV (note the apparent activation energy is *negative*) <br> $k$, Boltzmann's constant = $8.62 \times 10^{-5}$ eV/K <br> $T$, temperature (degrees Kelvin) |
| Thermomechanical Stress | Time to failure as a function of change in temperature | Stress generated by differing thermal expansion rates | $TF = B_0(T_0 - T)^{-n} \exp(E_a/kT)$ <br><br> where: <br><br> TF, time to failure <br> $B_0$, scale factor determined by experiment <br> $T_0$, stress-free temperature for metal (approximate metal deposition temperature for aluminum) <br> $N = 2\text{--}3$ |

(Continued)

| Table 2.5 (Continued) | | | |
| --- | --- | --- | --- |
| Model Name | Description/ Parameters | Application Examples | Model Equation |
| | | | $E_a$ = 0.5−0.6 eV for grain-boundary diffusion, ~1 eV for intragrain diffusion $k$, Boltzmann's constant = $8.62 \times 10^{-5}$ eV/K $T$, temperature (degrees Kelvin) |

For a known or suspected failure mechanism, all stimuli affecting the mechanism based on anticipated application conditions and material capabilities are identified. Typical stimuli may include temperature, electric field, humidity, thermomechanical stresses, vibration, and corrosive environments. The choice of acceleration test conditions is based on material properties and application requirements; the time interval should be reasonable for different purposes as well. Acceleration stressing must be weighed against generating failures that are not pertinent to the experiment, including those due to stress equipment or materials problems, or "false failures" caused by product overstress conditions that will never occur during actual product use. Once specific conditions are defined, a matrix of acceleration tests embodying these stimuli must be developed that allow separate modeling of each individual stimulus. A minimum of two acceleration test cells is required for each identified stimulus; three or more are preferred. Several iterations may be required in cases where some stimuli produce only secondary effects. For instance, high-temperature storage and thermal cycle stressing may yield similar failures, where high-temperature storage alone is the root stimulus.

Accelerating test sampling, defining failure criteria (i.e., the definition of a failure depends on the failure mechanism under evaluation and the test vehicle being used), and choosing an appropriate distribution (fitting the observed data to the proper failure distribution), then the acceleration factors are driven [9−11].

# CHAPTER 3

# Failure Mechanisms

It is a basic assumption that every failure has an explainable physical or chemical cause. This cause is called the failure mechanism. It is the goal of reliability physicists to identify the failure mechanism, determine what materials and stresses are involved, and quantify the results. An equation is derived that can be used to predict when failure will occur in the particular device under use conditions. The mechanisms described here are all well summarized in the JEDEC handbook JEP122G [3].

A single modern IC may have more than a billion transistors, miles of narrow metal interconnect ions, and billions of metal vias or contacts. As circuit entities, metal structures dominate the semiconductor transistors and their description is as challenging and necessary as that of semiconductors. Transistor oxides have shrunk, and dimensions are now on the order of 5 $SiO_2$ molecules thick. Metal and oxide materials failure modes are more significant in the deep-submicron technologies.

Besides, reliability tests are designed to reproduce failures of a product, which can occur in actual use. Understanding failure mechanisms from the results of reliability testing is extremely important to know the product is reliable in actual use. The effect of stresses (temperature, humidity, voltage, current, etc.) on the occurrence of failures can be identified by understanding the failure mechanisms. The product reliability in actual use can be predicted from the results of the reliability tests, which are conducted under acceleration conditions.

Failure mechanisms of electronic devices could be classified as those related to wafer, assembly, mounting, and handling processes.

## 3.1 TIME-DEPENDENT DIELECTRIC BREAKDOWN

Time-dependent dielectric breakdown (TDDB), also known as oxide breakdown, is one of the most important failure mechanisms in the semiconductor technology. TDDB refers to the wear-out process in gate oxide of the transistors; TDDB as well as electromigration and hot carrier effects are the main contributors to semiconductor device

Reliability Prediction from Burn-In Data Fit to Reliability Models. DOI: http://dx.doi.org/10.1016/B978-0-12-800747-1.00003-5

wear-out. However, the exact physical mechanisms leading to TDDB is still unknown [12−14].

It is generally believed that TDDB occurs as the result of defect generation by electron tunneling through gate oxide. When the defect density reaches a critical density, a conducting path would form in the gate oxide, which changes the dielectric properties. Defects within the dielectric, usually called traps, are neutral at first but quickly become charged (positively or negatively) depending to their locations in regard to anode and cathode.

Currently, the accepted model for oxide breakdown is the percolation model in which breakdown occurs when a conduction path is formed by randomly distributed defects generated during electrical stress. It is believed that when this conduction path is formed, the soft breakdown happens. When the sudden increase in conductance happens as well as a power dissipation rate more than a certain threshold, hard breakdown occurs. The whole process is very complicated but could be explained by physics of breakdown [15].

### 3.1.1 Early Models for Dielectric Breakdown

During an electric stress, the time to dielectric breakdown depends on electric stress parameters such as the electric field, temperature, and the total area of the dielectric film. From the 1970s until 2000, several models were suggested for the time to breakdown in dielectric layers.

The intrinsic failure, which occurs in defect-free oxide, is modeled in four forms:

- Bandgap ionization model happens in thick oxides; it occurs when the electron energy reaches the oxide bandgap and causes electron-hole pairs.
- Anode hole injection ($1/E$) model happens when the electron injected from the cathode gate gets enough energy to ionize the atoms and create hot holes; some of the holes tunnel back to the cathode and either create traps in the oxide or increase the cathode field and eventually the sudden oxide breakdown. The time to breakdown, $t_{BD}$, has a reciprocal electric field dependence ($1/E$) from the Fowler−Nordheim electron tunneling current:

$$t_{BD} = t_0 \exp\left(\frac{G}{E_{OX}}\right)\exp\left(\frac{E_a}{\kappa T}\right) \tag{3.1}$$

where $E_{OX}$ is the electric field across the oxide and $G$ and $t_0$ are constants.

- Thermochemical ($E$) model relates the defect generation to the electrical field. The applied field interacts with the dipoles and causes the oxygen vacancies and hence the oxide breakdown. The lifetime function has the following form:

$$t_{BD} = t_0 \exp(-\gamma E_{OX})\exp\left(\frac{E_a}{\kappa T}\right) \tag{3.2}$$

where $t_0$ and $\gamma$ are constants.

- Anode-hydrogen release model happens when electrons at anode release hydrogen which diffuses through the oxide and generate electron traps.

For thick dielectrics, the dielectric field is an important parameter controlling the breakdown process, while temperature dependence of dielectric breakdown is another key point. The $E$ and $1/E$ models can only fit part of the electric field, as shown in the following diagram (Figure 3.1). There are some articles trying to unify both models as well [16−18]. However, both of those models are not applicable to the ultrathin oxide layers. For ultrathin oxide layers (between 2 and 5 nm), other models are used.

### 3.1.2 Acceleration Factors

Acceleration factor could be given when a kind of stress similar to use stress applied to the device in a very short period of time results in

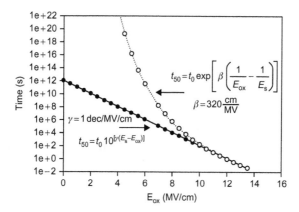

*Figure 3.1 Lifetime extrapolations based on the linear E and 1/E models shows the large discrepancies at the lower electric fields; both models are the same for electric field above 9 MV/cm for the field acceleration values [10].*

device failure. Failure may occur because of mechanical or electrical mechanisms. The device lifetime would change with the change of the operating parameters. "This change is quantified as the acceleration factor and is defined as the ratio of the measured failure rate of the device at one stress condition to the measured failure rate of identical devices stressed at another condition" [14]:

$$A_f = \frac{\text{MTTF}_{\text{use}}}{\text{MTTF}_{\text{stress}}} = \frac{t_{\text{BDuse}}}{t_{\text{BDstress}}} \tag{3.3}$$

where MTTF is mean time to failure of the device.

For thick dielectrics, the TDDB lifetime models are based on exponential law for field or voltage acceleration and Arrhenius law for the temperature.

The acceleration factor of E model TDDB is calculated by Ref. [14]:

$$A_{f\text{TDDB}} = \exp(\gamma(E_{\text{OX}} - E_{\text{OXstress}}))\exp\left(\frac{E_{\text{aTDDB}}}{k}\left(\frac{1}{T} - \frac{1}{T_{\text{stress}}}\right)\right) \tag{3.4}$$

Reliability engineering uses several oxide breakdown test configurations such as voltage ramp, current ramp, constant voltage stress test, and constant current stress test [19,20].

### 3.1.3 Models for Ultrathin Dielectric Breakdown

The dielectric thickness of modern semiconductor devices has been steadily thinning. The early models of thick dielectric breakdown should be revised to give the more accurate results for dielectric breakdown.

There are several models to describe the ultrathin dielectric breakdown both in devices and in the circuits. "Voltage-driven model for defect generation and breakdown" is a model that applies the percolation theory to find the breakdown in $SiO_2$ layers and suggests the use of a voltage-driven model instead of an electric field-driven one. It says that the tunneling electrons with energy related to the applied gate voltage are the driving force for defect generation and breakdown in ultrathin dielectrics. According to this model, despite the thick dielectric layer time to failure (TTF) breakdown, the ultrathin dielectric time to breakdown does not have the Arrhenius behavior anymore [13,21].

On the other hand, it is said that the exponential law with a constant voltage acceleration factor in ultrathin dielectrics would lead to two unphysical results:

1. projected lifetime for thinner oxides would be much longer than that of thick oxides at lower voltages;
2. the violation of the fundamental breakdown properties related to Poisson random statistics.

Therefore, the voltage dependence of time to breakdown follows a power law behavior instead of an exponential law. The interrelationship between voltage and temperature dependencies of oxide breakdown shows that the temperature activation with non-Arrhenius behavior is consistent with power law voltage dependence. This dependency for $t_{BD}$ has the following form:

$$t_{BD} \propto V_{Gate}^{-n} \tag{3.5}$$

where $n$ is the constant driven from the experimental data. The power law voltage dependence can eliminate the two unphysical results of the exponential law as well [22].

Other experimental results support the power law model, but give a wide range of numbers for $n$ for different gate voltages [23].

Further studies show that, while there is neither microscopic information about the breakdown defect and the interaction with stress voltage and temperature, nor a complete explanation of temperature dependence of dielectric breakdown of ultrathin dielectric layers, there are empirical relations between stress voltage and temperature:

$$\frac{V}{t_{BD}} \frac{\partial t_{BD}}{\partial V} = \text{const} = n(T) \tag{3.6}$$

which shows a power law dependence of $t_{BD}$ on voltage and covers experimental data in this regard and

$$\frac{d}{dT} \left( \frac{1}{t_{BD}} \frac{\partial t_{BD}}{\partial V} \right) \big|_{t_{BD}} = 0 \tag{3.7}$$

which shows the voltage acceleration at a fixed value of $t_{BD}$.

The third relation gives the $t_{BD}$ temperature dependence for a given voltage:

$$t_{BD} = t_{BD0}(V)\exp\left(\frac{a(V)}{T} + \frac{b(V)}{T^2}\right) \tag{3.8}$$

$\frac{b(V)}{T^2}$ in the above relation shows the non-Arrhenius behavior of temperature [24,25].

Further articles associate a linear relation to $n(T)$ for simplicity (which $n(T)$ less than 0):

$$n(T) = a' + b'T \tag{3.9}$$

Then the power law relationship between voltage and time to breakdown is:

$$t_{BD} \propto V_{GatetoSource}^{a'+b'T} \tag{3.10}$$

Since according to Weibull distribution, the TTF has the following relation with cumulative failure probability ($F$):

$$t_{BD} = \alpha\left[\ln\frac{1}{1-F}\right]^{\frac{1}{\beta}} \tag{3.11}$$

One can get the following approximation:

$$t_{BD} \propto F^{\frac{1}{\beta}} \tag{3.12}$$

Considering the lifetime dielectric breakdown as a function of gate oxide area as:

$$t_{BD} \propto \left(\frac{1}{WL}\right)^{\frac{1}{\beta}} \tag{3.13}$$

where $W$ is the channel width and $L$ is the channel length and taking the temperature dependence as:

$$t_{BD} \propto \exp\left(\frac{a(V)}{T} + \frac{b(V)}{T^2}\right) \tag{3.14}$$

the time-dependent breakdown becomes as a function of the following terms:

$$t_{BD} = A\left(\frac{1}{WL}\right)^{\frac{1}{\beta}} F^{\frac{1}{\beta}} V_{GatetoSource}^{a'+b'T} \exp\left(\frac{a(V)}{T} + \frac{b(V)}{T^2}\right) \tag{3.15}$$

where $\beta$, $a$, $b$, $a'$, and $b'$ can be driven according to the experimental data [26].

One can use generalized Weibull distribution to get the better adjustment of experimental and simulated data in comparison with the common Weibull one, which leads to the improved $t_{BD}$ extrapolation and area scaling [27].

### 3.1.4 Statistical Model

The statistics of gate oxide breakdown are usually described using the Weibull distribution:

$$F(t) = 1 - \exp[-(t/\alpha)^\beta] \tag{3.16}$$

where $\alpha$ is the scale parameter and $\beta$ is the shape parameter. Weibull distribution is an extreme-value distribution in $\ln(x)$ and is a "weakest link" type of problem. Here $F$ is the cumulative failure probability, $x$ can be either time or charge, $\alpha$ is the scale parameter, and $\beta$ is the shape parameter. The "weakest link" model was formulated by Sune et al. and described oxide breakdown and defect generation via a Poisson process [28]. In this model, a capacitor is divided into a large number of small cells. It is assumed that during oxide stressing neutral electron traps are generated at random positions on the capacitor area. The number of traps in each cell is counted, and at the moment that the number of traps in one cell reaches a critical value, breakdown will occur. Dumin [29] incorporated this model to describe failure distributions in thin oxides.

The Weibull slope $\beta$ is an important parameter for reliability projections. A key advance was the realization that $\beta$ is a function of oxide thickness $t_{ox}$, becoming smaller as $t_{ox}$ decreases. The smaller $\beta$ for thinner oxide is explained as the conductive path in the thinnest oxides consists of only a few traps and therefore has a larger statistical spread. The shape parameter's oxide thickness dependence is shown in Figure 4.1. The dependence of $\beta$ on $t_{ox}$ can be fitted to:

$$\beta = (t_{int} + t_{ox})/a_0 \tag{3.17}$$

where $a_0$ is the linear defect size with a fitted value of 1.83 n, and $t_{int}$ is the interfacial layer thickness with a fitted value of 0.37 nm [30]. It can be found that $\beta$ is approaching one as $T_{ox}$ is near 1 nm, which means Weibull distribution will become exponential distribution. Lognormal distribution has also been used to analyze acceleration test data of dielectric breakdown. Although it may fit failure data over a limited sample set, it has been demonstrated that the Weibull distribution more accurately fits large samples of TDDB failures. An important disadvantage of lognormal distribution is that it does not predict the observed area dependence of TBD for ultrathin gate oxides and the distribution itself does not fit into a multiple mechanism system that is linearly combined.

## 3.2 HOT CARRIER INJECTION

Hot carriers in the semiconductor device are the cause of a distinct wearout mechanism, the hot carrier injection (HCI). Hot carriers are produced when the source-drain current flowing through the channel attains high energy beyond the lattice temperature. Some of these hot carriers gain enough energy to be injected into the gate oxide, resulting in charge trap and interface state generation. The latter may lead to shifts in the performance characteristics of the device, e.g., the threshold voltage, transconductance, or saturation current, and eventually to its degradation. The rate of HCI is directly related to the channel length, oxide thickness, and the operating voltage of the device. Since the latter are minimized for optimal performance, the scaling has not kept pace with the reduction in channel length. Current densities have been increased with a corresponding increase in device susceptibility to hot carrier effects.

Hot carriers are generated during the operation of semiconductor devices as they switch states. As carriers travel through the channel from source to drain, the lateral electric field near the drain junction causes carriers to become hot. A small number of these hot carriers gain sufficient energy—higher than the $Si-SiO_2$ energy barrier of about 3.7 eV—to be injected into the gate oxide. In negative-channel metal-oxide semiconductor (NMOS) devices, hot electrons are generated while hot holes are produced in positive-channel metal-oxide semiconductor (PMOS) devices. Injection of either carrier results in three primary types of damage: trapping of electrons or holes in preexisting traps, generation of new traps, and generation of interface traps. These traps may be classified by location while their effects vary.

Interface traps are located at or near the $Si-SiO_2$ interface and directly affect transconductance, leakage current, and noise level. Oxide traps are located farther away from the interface and affect the long-term MOSFET stability, specifically the threshold voltage. Effects of defect generation include threshold voltage shifts, transconductance degradation, and drain current reduction. Negative bias temperature instability (NBTI) seems to have similar degradation patterns, except for PMOS.

Hu et al. proposed the "lucky" electron model for hot carrier effects. This is a probabilistic model proposing that a carrier must first gain enough kinetic energy to become "hot," and then the carrier momentum must become redirected perpendicularly so the carrier can enter the oxide. The current across the gate is denoted by $I_{gate}$ and during normal operation its value is negligible. Degradation due to hot carriers is proportional to $I_{gate}$, making the latter a good monitor of the former. As electrons flow in the channel, some scattering of the electrons in the lattice of the silicon substrate occurs due to interface states and fixed charges (interface defects). As electron scattering increases, the mobility of the hot carriers is reduced, thereby reducing the current flowing through the channel. Over a period of time, hot carriers degrade the silicon bonds with an attendant increase in electron scattering due to an increase in interface and bulk defects. As a result, the transistor slows down over a period of time. The hot carrier lifetime constraints in an NMOS device limit the current drive that can be used in a given technology. By improving the hot carrier lifetime, the current drive can be increased, thereby increasing the operating speed of a device, such as a microprocessor [31–36].

### 3.2.1 Hot Carrier Effects

HCI is the second major oxide failure mechanism that occurs when the transistor electric field at the drain-to-channel depletion region is too high. This leads to the HCI effects that can change circuit timing and high-frequency performance. HCI rarely leads to catastrophic failure but the typical parameters affected are: $I_{Dsat}$, $G_m$, and $V_{th}$, weak inversion subthreshold slope and increased gate-induced drain leakage.

HCI happens if the power supply voltage is higher than needed for the design, the effective channel lengths are too short, there is a poor oxide interface or poorly designed drain substrate junction, or overvoltage accidentally occurs on the power rail. The horizontal electric field in the channel gives kinetic energy to the free electrons moving from the inverted portion of the channel to the drain; when the kinetic energy is high enough, electrons strike Si atoms around the drain substrate interface, causing impact ionization. Electron-hole pairs are produced in the drain region and scattered. Some carriers go into the substrate, causing an increase in substrate current, and the small fraction has enough energy to cross the oxide barrier and cause damage. A possible mechanism is that a hot electron breaks a hydrogen silicon bond at the Si-SiO$_2$ interface. If the Si and hydrogen recombine, then no interface trap is created. If the hydrogen diffuses away, then an interface trap is created.

Once the hot carrier enters the oxide, the vertical oxide field determines how deeply the charge will go. If the drain voltage is positive with respect to the gate voltage, then holes entering the oxide near the drain are accelerated deeper into the oxide and the electrons in the same region will be retarded from living the oxide interface. Electric field in the channel restricts the damage to oxide over drain substrate depletion region, with only a small amount of damage, just outside the depletion layer [37].

In submicron range electronic devices, one of the major reliability problems is hot carrier degradation. This problem is related to the continuous increase of the electrical fields in both oxide and silicon. Under the influence of the high lateral fields in short channel MOSFETs, electrons and holes in the channel and pinch off regions of the transistor can gain sufficient energy to overcome the energy barrier or tunnel into the oxide. This leads to injection of a gate current into the oxide, and subsequently to the generation of traps, both at the interface and in the oxide, and to electron and hole trapping in the oxide, which will cause changes in transconductance, threshold voltage, and drive currents of the MOSFET [38,39].

### 3.2.2 Acceleration Factor

Hot carrier phenomena are accelerated by low temperature, mainly because this condition reduces charge de-trapping. A simple acceleration model for hot carrier effects is as follows:

$$\text{AF} = t_{50(2)}/t_{50(1)} \tag{3.18}$$

$$\text{AF} = \exp([E_a/k][1/T1 - 1/T2] + C[V2 - V1]) \tag{3.19}$$

where:

AF = acceleration factor of the mechanism;

$t_{50(1)}$ = rate at which the hot carrier effects occur under conditions $V1$ and $T1$;

$t_{50(2)}$ = rate at which the hot carrier effects occur under conditions $V2$ and $T2$;

$V1$ and $V2$ = applied voltages for $R1$ and $R2$, respectively;

$T1$ and $T2$ = applied temperatures (degrees K), respectively;

$E_a$ = $-0.2$ eV to $-0.06$ eV; and

$C$ = a constant [40].

Typically the assessment of devices with regard to hot carrier effect will take place at low temperatures ($-20$ to $-70°C$) due to the negative activating energy associated with this mechanism.

Temperature acceleration is often treated as a minor effect in most HCI models; however, in order to consider possible large temperature excursion, FaRBS includes temperature acceleration effect based on the HCI lifetime model as given below [35]:

$$t_f = B(I_{Sub})^{-n} \exp(E_a/kT) \text{ for } n\text{-type} \qquad (3.20)$$

$$t_f = B(I_{Gate})^{-m} \exp(E_a/kT) \text{ for } p\text{-type} \qquad (3.21)$$

The combination of temperature effect produces a more complicated HCI lifetime model as below [23]:

$$t_f = A_{HCI}\left(\frac{I_{sub}}{W}\right)^{-n} \exp\left(\frac{E_{aHCI}}{kT}\right) \qquad (3.22)$$

Then it should be possible to calculate the acceleration factor from the above equation.

### 3.2.3 Statistical Models for HCI Lifetime

There is little discussion in literature about a proper statistical lifetime distribution model for HCI. However, Weibull failure statistics [31,32] for failure rate and lognormal distribution for HCI lifetime [41] are reported as well. A logical hypothesis for the lifetime distribution would be the exponential one. This is a good assumption because as a device becomes more complex, with millions of gates, it may be considered as a system. The failure probability of each individual gate is not most likely an exponential distribution. However, the cumulative effect of early failures and process variability, ensuring each gate has a different failure rate, widens the spread of the device failures. The end result is that intrinsic HCI becomes statistically more random as the failures occur at a constant rate.

The effect of process variation on hot carrier reliability characteristics of MOSFET is another field of research, as well. A significant variation of hot carrier lifetime across the wafers due to gate length variation is reported. Since the nonuniformity of gate length is always present, it is necessary to consider enough margin of hot carrier lifetime and the accurate evaluation of gate length variation for optimizing device performance should be considered [32].

### 3.2.4 Lifetime Sensitivity

HCI lifetime is sensitive to changes in the input parameters. The acceleration factor for HCI is:

$$A_{F,\text{HCD}} = \exp(B(1/V_{\text{dd}} - 1/V_{\text{dd,max}}))\tag{3.23}$$

HCI continues to be a reliability concern as device feature sizes shrink. HCI is a function of internal electric fields in the device and as such is affected by channel length, oxide thickness, and device operating voltage. Shorter channel lengths decrease reliability but the oxide thickness and the voltage may also be reduced to help alleviate the reduction in reliability. Another way of improving hot carrier reliability may be by shifting the position of the maximum drain so it is deeper in the channel. This would result in hot carriers being generated farther away from the gate and Si-SiO2 interface, reducing the likelihood of injection into the gate. Another method is to reduce the substrate current by using a lightly doped drain (LDD) where part of the voltage drop is across an LDD extension not covered by the gate. Annealing the oxides in $NH_3$, $N_2O$, or NO or growing them directly in $N_2O$ or NO improves their resistance to interface state generation by the hot carrier.

## 3.3 NEGATIVE BIAS TEMPERATURE INSTABILITY

NBTI happens to PMOS devices under negative gate voltages at elevated temperatures. The degradation of device performance, mainly manifested as the absolute threshold voltage $V_{\text{th}}$ increase and mobility, transconductance and drain current $I_{\text{dsat}}$ decrease, is a big reliability concern for today's ultrathin gate oxide devices [42]. Deal [43] named it "Drift VI" and discussed the origin in the study of oxide surface charges. Goetzberger et al. [44] investigated surface state change under combined bias and temperature stress through experiments that utilized MOS structures formed by a variety of oxidizing, annealing, and metalizing procedures. They found an interface trap density $D_{it}$ peak in the lower half of the bandgap and p-type substrates gave higher $D_{it}$ than n-type substrates. The higher the initial $D_{it}$, the higher the final stress-induced $D_{it}$. Jeppson et al. [45] first proposed a physical model to explain the surface trap growth of MOS devices subjected to negative bias stress. The surface trap growth was described as diffusion controlled at low fields and tunneling limited at height fields. The power law relationship ($t^{1/4}$) was also proposed for the first time. The study of NBTI has been very active in recent years since the interface trap density induced by NBTI increases with decreasing oxide thickness, which means NBTI is more severe to ultrathin oxide devices. New developments of NBTI modeling and surface trap analysis have been reported in recent years. At the same time, effects of various process parameters on NBTI had been studied in order to minimize the NBTI. Schroder et al. [46] reviewed pre-2003 experimental results and various proposed physical models together with the effects of manufacturing process parameters. Detailed latest reviews can also be found in Refs. [47−51]. In this section, the up-to-date research discoveries of NBTI failure mechanism, models, and related parameters will be briefly discussed.

### 3.3.1 Degradation Models

The time dependence of the threshold voltage shift $\Delta V_{th}$ is found to follow a power law model:

$$\Delta V_{th}(t) = At^n \tag{3.24}$$

where A is a constant, which depends on oxide thickness, field, and temperature. The time exponent $n$ is a sensitive measure of the diffusion species. The theoretical value of the exponent parameter $n$ is 0.25 according to the solution of diffusion equations [52]. Chakravarthi [53] suggested that $n$ varies around 0.165, 0.25, and 0.5 depending on the reaction process and the type of diffusion species. According to Alam et al. [51], $n = 1 = 2$ for proton, $n = 1 = 6$ for molecular H2, and $n = 1 = 4$ for atomic H. $n$ was also reported to change from $\sim 0{:}25$ initially (stress time $\sim 100$ s) to 0.16 at $10^6$ s stress time [49]. NBTI degradation is thermally activated and sensitive to temperature. The temperature dependence of NBTI is modeled by the Arrhenius relationship. The activation energy appears to be highly sensitive to the types of potential reacting species and to the type of oxidation methods used [46]. Reported activation energies range from 0.18 to 0.84 eV [54,55].

Improved models have been proposed after the simple power law model. Considered the temperature and gate voltage, $\Delta V_{th}$ can be expressed as:

$$\Delta V_{th}(t) = B \exp(\beta V_g) \exp(-E_a/kT) t^{0.25} \quad (3.25)$$

where $B$ and $\beta$ are constants and $V_g$ is the applied gate voltage. Considering the effects of gate voltage and oxide field, Mahapatra et al. [56] proposed a first order $N_{it}$ model.

$$\Delta N_{it}(t) = K(C_{ox}(V_g - V_{th}))^{0.5} \exp(\beta E_{ox}) \exp(-E_a/kT) t^{0.25} \quad (3.26)$$

where $K$ is a constant.

### 3.3.2 Lifetime Models

NBTI failure is defined as $\Delta V_{th}$ reaches a threshold value. Based on the degradation models such as previous equations, NBTI lifetime can be represented as:

1. Field model:

$$\tau = C_1 E_{ox}^{-n} \exp(E_a/kT) \quad (3.27)$$

   where $C_1$ is a constant.
2. Voltage model:

$$\tau = C_2 \exp(-\beta V_g) \exp(E_a/kT) \quad (3.28)$$

   where $C_2$ is a constant.

## 3.4 ELECTROMIGRATION

Electrons passing through a conductor transfer some of their momentum to its atoms. At sufficiently high electron current densities (greater than $10^5$ A/cm$^2$ [57]), atoms may shift toward the anode side. The material depletion at the cathode side causes circuit damage due to decreased electrical conductance and eventual formation of open circuit conditions. This is caused by voids and micro-cracks, which may increase the conductor resistance as the cross-sectional area is reduced. Increased resistance alone may result in device failure, yet the resulting increase in local current density and temperature may lead to thermal runaway and catastrophic failure [58], such as an open circuit failure. Alternatively, short circuit conditions may develop due to excess material buildup at the anode. Hillocks form where there is excess material, breaking the oxide layer, allowing the conductor to come in contact with other device features. Other types of damage include whiskers, thinning, localized heating, and cracking of the passivation and interlevel dielectrics [59].

This diffusive process, known as electromigration, is still a major reliability concern despite vast scientific research as well as electrical and materials engineering efforts. Electromigration can occur in any metal when high current densities are present. In particular, the areas of greatest concern are the thin-film metallic interconnects between device features, contacts and vias [59].

### 3.4.1 Lifetime Prediction

Modeling electromigration mean time to failure (MTTF) from the first principles of the failure mechanism is difficult. While there are many competing models attempting to predict time to failure from first principles, there is no universally accepted model.

Currently, the favored method to predict TTF is an approximate statistical one given by *Black's* equation, which describes the MTTF by:

$$MTTF = A(j_e)^{-n} \exp(E_a/kT) \qquad (3.29)$$

where $j_e$ is the current density and $E_a$ is the EM activation energy. Failure times are described by the lognormal distribution [60]. The symbol $A$ is a constant, which depends on a number of factors, including grain size, line structure and geometry, test conditions, current density, thermal history, etc. Black determined the value of $n$ to equal 2. However, $n$ is highly dependent on residual stress and current density [8 m] and its value is highly controversial.

A range of values for the EM activation energy, $E_a$, of aluminum (Al) and aluminum alloys is also reported. The typical value is $E_a = 0.6 \pm 0.1$ eV. The activation energy can vary due to mechanical stresses caused by thermal expansion. Introduction of 0.5% Cu in Al interconnects may result in $n = 2.63$ and an activation energy of $E_a = 0.95$ eV. For multilevel Damascene Cu interconnects, an activation energy of $E_a = 0.94 \pm 0.11$ eV at a 95% confidence interval (CI) and a value of the current density exponent of $n = 2.03 \pm 0.21$ (95% CI) were found [61].

### 3.4.2 Lifetime Distribution Model

Traditionally, the EM lifetime has been modeled by the lognormal distribution. Most test data appear to fit the lognormal distribution, but these data are typically for the failure time of a single conductor [62].

Through the testing of over 75,000 Al(Cu) connectors, Gall et al. [62] showed that the electromigration failure mechanism does follow the lognormal distribution. This is valid for the TTF of the first link with the assumption that the first link failure will result in device failure. The limitation is that a lognormal distribution is not scalable. A device with different numbers of links will fail with a different lognormal distribution. Thus, a measured failure distribution is valid only for the device on which it is measured. Gall et al. also showed that the Weibull (and thus the exponential) distribution is not a valid model for electromigration. Even though the lognormal distribution is the best fit for predicting the failure of an individual device due to EM, the exponential model is still applicable for modeling EM failure in a system of many devices where the reliability is determined by the first failure of the system.

### 3.4.3 Lifetime Sensitivity

The sensitivity of the electromigration lifetime can be observed by plotting the lifetime as a function of the input parameters. For EM, the most significant input parameters corresponding to lifetime are the temperature ($T$) and current density ($j_e$). The lifetime may be normalized using an acceleration factor. Substituting Black's equation and assuming an exponential failure distribution into

$$A_f = \lambda_{\text{rated}}/\lambda \tag{3.30}$$

provides the acceleration factor for EM,

$$A_{f,EM} = (j_e/j_{e,\text{rated}})^{-n}\exp[(E_{a,EM}/kT)(1/T - 1/T_{\text{rated}})] \tag{3.31}$$

Obviously, $T$ has a much greater impact on $A_f$ than $j_e$.

As device features continue to shrink and interconnect current densities grow, EM will remain a concern. New technologies may reduce the EM impact of increasing densities but new performance requirements emerge that require increased interconnect reliability under conditions of decreased metallization inherent reliability [64]. Thus, EM will remain a design and wearout issue in future semiconductor designs.

## 3.5 SOFT ERRORS DUE TO MEMORY ALPHA PARTICLES

One of the problems which hinder development of larger memory sizes or the miniaturization of memory cells is the occurrence of soft errors due to alpha particles. This phenomenon was first described by T. C. May. U (uranium) and Th (thorium) are contained in very low concentrations in package materials and emit alpha particles that enter the memory chip and generate a large concentration of electron-hole pairs in the silicon substrate. This causes a change in the electric potential distribution of the memory device amounting to electrical noise, which, in turn can cause changes in the stored information. Inversion of

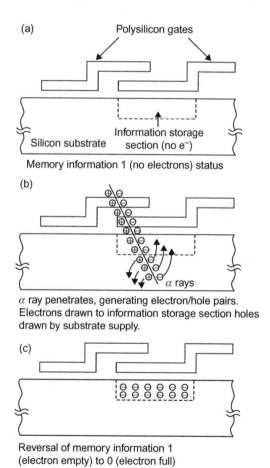

Figure 3.2 Memory cell model of soft error.

memory information is shown in the figure. The generated holes are pulled toward the substrate with its applied negative potential. Conversely, electrons are pulled to the data storage node with its applied positive potential. A dynamic memory filled with charge has a data value of 0. An empty or discharged cell has a value of 1. Therefore, a data change of 1-0 occurs when electrons collect in the data storage node. Such a malfunction is called "memory cell model" of a soft error (Figure 3.2).

The "bit line model" occurs due to change of the bit line electric potential. The bit line's electric potential varies with the data of the memory cell during readout, and is compared with the reference potential, resulting in a data value of 1 or 0. A sense amplifier is used to amplify the minute amount of change. If $\alpha$ particles penetrate the area near the bit path during the minimal time between memory readout and sense amplification, the bit path potential changes. An information 1-0 operation error results when the bit path potential falls below the reference potential. Conversely, if the reference potential side drops, an information 0-1 operation error results. The memory cell model applies only to information 1-0 reversal, while the bit path model covers both information 1-0 and 0-1 reversals. The generation rate of the memory cell model is independent of memory cycle time because memory cell data turns over. Since the bit path model describes problems that occur only when the bit line is floating after data readout, increased frequency of data readout increases the potential for soft errors, i.e., the bit path model occurrence rate is inversely proportional to the cycle time. In product, the "mixed model" combined model describes the combination of the memory cell and bit path models.

# CHAPTER 4

# New M-HTOL Approach

Microelectronics integration density is limited by the reliability of the manufactured product at a desired circuit density. Design rules, operating voltage, and maximum switching speeds are chosen to insure functional operation over the intended lifetime of the product. Thus, in order to determine the ultimate performance for a given set of design constraints, the reliability must be modeled for its specific operating condition.

Reliability modeling for the purpose of lifetime prediction is therefore the ultimate task of a failure physics evaluation. Unfortunately, all the industrial approaches to reliability evaluation fall short of predicting failure rates or wear-out lifetime of semiconductor products. This is attributed mainly to two reasons: the lack of a unified approach for predicting device failure rates and the fact that all commercial reliability evaluation methods rely on the acceleration of one, dominant, failure mechanism.

Over the last several decades, our knowledge about the root cause and physical behavior of the critical failure mechanisms in microelectronic devices has grown significantly. Confidence in the reliability models has lead to more aggressive design rules that have been successfully applied to the latest VLSI technology. One result of improved reliability modeling has been acceleration performance, beyond the expectation of Moore's law. A consequence of more aggressive design rules has been a reduction in the weight of a single failure mechanism. Hence in modern devices, there is no single failure mode that is more likely to occur than any other as guaranteed by the integration of modern failure physics modeling and advanced simulation tools in the design process. The consequence of more advanced reliability modeling tools is a new phenomenon of device failures resulting from a combination of several competing failure mechanism. Hence, a new approach is required for reliability modeling and prediction.

Multiple failure mechanism physics together with a novel testing method is suggested to overcome the inaccurate results of "zero

Reliability Prediction from Burn-In Data Fit to Reliability Models. DOI: http://dx.doi.org/10.1016/B978-0-12-800747-1.00004-7

failure" reported data. "Zero failure" criteria has evolved due to the legal requirement of companies to not show that they tolerate "failure." This has made the actual prediction of a potential failure from a test that generated zero failures impossible. Hence, we must replace zero failures with either degradation extrapolation or with actual failure generation at conditions which are actually accelerating failures to occur. This allows us to eliminate the false statistical certainty deriving from the zero failure criterion, forcing a meaningless "60%" confidence on the statistic. We furthermore reduce the uncertainty of the acceleration factor (AF) by combining known or published models for exactly the failure mechanisms we are interested to test.

Our linear matrix method depends on the further statistical assumption that each mechanism, although statistically independent, will combine approximately linearly in a large enough system with millions of transistors and billions of connections and wires. This allows the application of the sum-of-failure-rate approach that is demanded by the JEDEC standard on which this system is based. Hence, our multiple failure mechanism assumption, together with multiple acceleration testing and the assumption of constant failure rate (CFR), approximates the interaction and relative importance of failures in electronic systems (consisting of millions of components). These assumptions form the basis of the proposed approach.

## 4.1 PROBLEMATIC ZERO FAILURE CRITERIA

One of the fundamentals of understanding a product's reliability requires an understanding of the calculation of the failure rate. The calculation of failure rates is an important metric in assessing the reliability performance of a product or process. This data can be used as a benchmark for future performance or an assessment of past performance, which might signal a need for product or process improvement. Reliability data are expressed in numerous units of measurement. However, most documents use the units of FITs (failures in time), where one FIT is equal to one failure occurring in $10^9$ device hours. More literally, FIT is the number of failures per $10^9$ device hours. It is typically used to express the failure rate. Similarly, it can be defined as 1 ppm per 1000 hours of operation or one failure per 1000 devices run for 1 million hours of operation.

The traditional method of determining a product's failure rate is through the use of acceleration high temperature operating life (HTOL) tests performed on a sample of devices randomly selected from its parent population. The failure rate obtained on the life test sample is then extrapolated to end-use conditions by means of predetermined statistical models to give an estimate of the failure rate in the field application. Although there are many other stress methods employed by semiconductor manufacturers to fully characterize a product's reliability, the data generated from operating life test sampling is the principal method used by the industry for estimating the failure rate of a semiconductor device in field service.

The HTOL or steady state life test is performed to determine the reliability of devices under operation at high temperature conditions over an extended period of time. It consists of subjecting the parts to a specified bias or electrical stressing, for a specified amount of time, and at a specified high temperature. Unlike a production burn-in test, which accelerates early life failures, HTOL testing is applied to assess the potential operating lifetimes of the sample population. It is therefore more concerned with acceleration of wear-out failures. As such, life tests should have sufficient durations to assure that the results are not due to early life failures or infant mortality. The HTOL qualification test is usually performed as the final qualification step of a semiconductor manufacturing process. The test consists of stressing some number of parts, usually about 100, for an extended time, usually 1000 h, at an acceleration voltage and temperature. Two features shed doubt on the accuracy of this procedure [65].

In many cases, the individual components in an engineering system may have suffered no or perhaps only a single failure during their operating history. Such a history may be that of continuous operation over a period of time or it may be over a number of discrete demands. Probabilistic safety assessment provides a framework to analyze the safety of large industrial complexes. Therefore, the lack of data has encouraged either the elicitation of opinion from experts in the field (i.e., "expert judgment technique") or making a confidence interval. Several articles show the history of failure rate estimation for low probability events and the means to handle the problem [66–68].

In general, apart from Bayesian approach, when a reliability test ends in zero units having failed, traditionally reliability calculations suggest that the estimated failure rate is also zero, assuming an exponential distribution. However, obviously this is not a realistic estimate of a failure rate, as it does not take into account the number of units. In such cases, the first approximation is to select a failure rate that makes the likelihood of observing zero failures equal to 50% [69–71]. In other words, a failure rate that carries a high probability of observing zero failures for a given reliability test is selected. An upper 100 $(1 - \alpha)$ confidence limit for $\lambda$ is given by

$$\lambda_{100(1-\alpha)} = \frac{\chi^2_{2;100(1-\alpha)}}{2nT} = \frac{-\ln \alpha}{nT} \tag{4.1}$$

where $n$ is the total number of devices, $T$ is the fixed time end of test, and $\chi^2_{2;100(1-\alpha)}$ is the upper $100(1 - \alpha)$ percentile of the chi-square distribution with 2 degrees of freedom. In practice, $\alpha$ is often set to 0.5, which is referred to as 50% zero failures estimate of $\lambda$. However, $\alpha$ can theoretically be set to any probability value desired. $\lambda_{50}$ is the failure rate estimate that makes the likelihood of observing zero failures in a given reliability test equal to 50%.

This is the basis of calculating the semiconductor failure rates. In the case of semiconductor devices, it is said that since failure rates calculated from actual HTOL test data are not the result of predictions (such as those contained in MIL-HDBK-217), they are calculated from the number of device hours of test, the number of failures (if any), and the chi-square statistic at the 60% confidence level. The formula for failure rate calculation at a given set of conditions is as follows:

$$FR = \frac{\chi^2_{(2C+2)}}{2nT} \times 10^9 \text{ h} \tag{4.2}$$

where $\chi^2_{(2C+2)}$ is chi-square distribution factor with $2C + 2$ degrees of freedom (taken from chi-square tables), $C$ is total number of failures, $n$ is total number of devices tested, and $T$ is test duration for each device at the given conditions.

$$MTBF = \frac{10^9}{FR} \tag{4.3}$$

The calculation of the failure rates and MTBF (mean time before failure) values at a given condition is accomplished through the determination of the accelerating factor for that condition [73].

The problem is that the "zero failure rates" criterion is based on the inaccurate (and even incorrect) assumption of "single failure mechanism," on one hand, and the "confidence interval" which is built upon the mere zero (or at most) one data point, on the other hand. Unfortunately, with zero failures no statistical data is acquired. The other feature is the calculation of the AF. If the qualification test results in zero failures, it allows the assumption (with only 60% confidence!) that no more than half a failure occurred during the acceleration test. This would result, based on the example parameters, in a reported FIT = 5000/AF, which can be almost any value from less than 1 FIT to more than 500 FIT, depending on the conditions and model used for the voltage and temperature acceleration.

The accepted approach for measuring FIT would, in theory, be reasonably correct if there is only a single dominant failure mechanism that is excited equally by either voltage or temperature. For example, electromigration (described in Section 3.4) is known to follow Black's equation (described later) and is acceleration by increased stress current in a wire or by increased temperature of the device. If, however, multiple failure mechanisms are responsible for device failures, each failure mechanism should be modeled as an individual "element" in the system and the component survival is modeled as the survival probability of all the "elements" as a function of time.

If multiple failure mechanisms, instead of a single mechanism, are assumed to be time independent and independent of each other, FIT (CFR approximation) should be a reasonable approximation for realistic field failures. Under the assumption of multiple failure mechanisms, each will be acceleration differently depending on the physics that is responsible for each mechanism. If, however, an HTOL test is performed at an arbitrary voltage and temperature for acceleration based only on a single failure mechanism, then only that mechanism will be acceleration. In that instance, which is generally true for most devices, the reported FIT (especially one based on zero failures) will be meaningless with respect to other failure mechanisms.

Table 4.1 presents definitions of some of the terms used to describe the failure rate of semiconductor devices [72].

| Table 4.1 Definitions of Some of the Terms Used to Describe the Failure Rate of Semiconductor Devices | |
|---|---|
| Terms | Definitions/Descriptions |
| Failure rate ($\lambda$) | Measure of failure per unit of time. The useful life failure rate is based on the exponential life distribution. The failure rate typically decreases slightly over early life, then stabilizes until wear-out which shows an increasing failure rate (IFR). This should occur beyond useful life |
| Failure in time (FIT) | Measure of failure rate in $10^9$ device hours; e.g., 1 FIT = 1 failure in $10^9$ device hours |
| Total device hours (TDH) | The summation of the number of units in operation multiplied by the time of operation |
| Mean time to failure (MTTF) | Mean of the life distribution for the population of devices under operation or expected lifetime of an individual, MTTF = $1/\lambda$, which is the time where 63.2% of the population has failed. Example: for $\lambda$ = 10 FITs, MTTF = $1/\lambda$ = 100 million hours |
| Confidence level or limit (CL) | Probability level at which population failure rate estimates are derived from sample life test. The upper confidence level interval is used |
| Acceleration factor (AF) | A constant derived from experimental data which relates the times to failure at two different stresses. The AF allows extrapoltaion of failure rates from accelerated test conditions to use conditions |

## 4.2 SINGLE VERSUS MULTIPLE COMPETING MECHANISMS

Acceleration stress testing has been recognized to be a necessary activity to ensure the reliability of high-reliability electronics. The application of enhanced stresses is usually for the purpose of (i) ruggedizing the design and manufacturing process of the package through systematic step-stress and increasing the stress margins by corrective action (reliability enhancement testing); (ii) conducting highly compressed/acceleration life tests in the laboratory to verify in-service reliability (acceleration life tests); and (iii) eliminating weak or defective populations from the main population (screening or infant mortality reduction testing).

In general, acceleration life testing techniques provide a shortcut method to investigate the reliability of electronic devices with respect to certain dominant failure mechanisms occurring under normal operating conditions. Acceleration tests are usually planned on the assumption that there is a single dominant failure mechanism for a given device. However, the failure mechanisms that are dormant under normal use condition may start contributing to device failure under acceleration conditions and the life test data obtained from the acceleration test would be unrepresentative of the actual situation. Moreover, an acceleration stress accelerates various failure mechanisms simultaneously and the dominant failure mechanism is the one which gives the shortest predicate life [73].

A model which accommodates multiple failure mechanisms considers a system with $k$ failure mechanisms, each of which is independent of others and follows exponential distribution. The lifetime of such a system is the smallest of $k$ failure mechanism lifetimes. If the variable $X$ denotes the system lifetime, then

$$X = \min(X_1, X_2, \ldots, X_k) \tag{4.4}$$

This model is often called competing model. The probability of a system surviving failure type I at time $t$ is

$$R_i(t) = P(X_i > t) = 1 - G_i(t) \tag{4.5}$$

where $G_i(t)$ is the distribution function of lifetime for failure type $i$. From the assumption that the failure mechanism develops independently of one another, the probability of surviving all $k$ failure mechanisms at time $t$ is

$$P(X_i > t | i = 1, 2, \ldots, k) = \prod_{i=1}^{k}(1 - G_i(t)) \tag{4.6}$$

Then the distribution function of system failure is

$$F(t) = 1 - \prod_{i=1}^{k}(1 - G_i(t)) \tag{4.7}$$

Consider each failure mechanism as an "element" in the system. The system survives only if all the "elements" survive, just like a series system reliability model. Then the probability of survival of the system is

$$R(t) = R_1(t)R_2(t)\ldots R_k(t) \tag{4.8}$$

If each failure mechanism has a time-variant reliability distribution, then the system reliability distribution is also time dependant and rather complex. However, the simulations show that the exponential distribution for each failure distribution would result in reasonable approximation. Applying CFR to the assumptions, the system reliability distribution also follows the exponential distribution. For each failure mechanism:

$$R_i(t) = \exp(\lambda_i t) \tag{4.9}$$

Then

$$R(t) = \prod_{i=1}^{k} \exp(\lambda_i t) = \exp(\lambda t) \tag{4.10}$$

where $\lambda = \lambda_1 + \lambda_2 + \cdots + \lambda_k$ is the sum of all failure rates. The above competing model provides the basis to find the more accurate form of AFs of complex systems with multiple failure mechanisms.

In modern devices, there is no single failure mode that is more likely to occur than any other as guaranteed by the integration of modern failure and modern simulation tools in the design process. The consequence of more advanced reliability modeling tools is a new phenomenon of device failures resulting from a combination of several competing failure mechanisms. It seems that the multiple failure mechanism realm is going to find its way even to standard handbooks; JEDEC Standard, JESD85, gives the instruction to calculate multiple activation energy procedures for CFR distributions; it is supposed that the devices failed due to several different failure mechanisms to which can be assigned the appropriate activation energies [74,75].

## 4.3 AF CALCULATION

In this section, semiconductor voltage and temperature acceleration factors have been studied and modified in detail. The key factor is that the Arrhenius relationship (proposed to model the single failure mechanism effect in acceleration tests and applied to predict the system reliability even for multiple failure mechanisms by manufacturers) is not appropriate for the multiple failure mechanism criterion. Thereafter, a new model is proposed to estimate the system reliability. Unlike the single-mechanism realm, multiple failure mechanism suggests a method to separate and detect different failure mechanisms. The result could be summarized in saying that TDDB (time-dependent dielectric breakdown) dominates under high temperature with high voltage, HCI (hot carrier injection) dominates under high voltage and low temperature, and EM dominates under high temperature with low temperature. So the basic idea is that different mechanisms activate under different kinds of conditions. This section compares the results obtained under single failure mechanism assumption and those obtained under the multiple failure mechanism realm; in the first step, the failure rate of three failure mechanisms (EM, TDDB, and HCI) is calculated under the single-mechanism assumption and then they are recalculated under multiple failure mechanism condition; the results would be compared to those of use condition.

Generally, for multiple mechanisms, the system acceleration factor can be expressed based on those of the mechanisms. For $n$ independent mechanisms, it can be written as

$$\frac{\lambda_{s\text{-test}}}{AF_s} = \lambda_{s\text{-use}} = \sum_{i=1}^{n} \lambda_{i\text{-use}} \tag{4.11}$$

Then

$$\frac{\lambda_{s\text{-test}}}{AF_s} = \sum_{i=1}^{n} \frac{\lambda_{i\text{-test}}}{AF_i} \tag{4.12}$$

$$\frac{\lambda_{s\text{-test}}}{AF_s} = \sum_{i=1}^{n} \frac{\alpha_i \lambda_{s\text{-test}}}{AF_i} \tag{4.13}$$

$$\frac{1}{AF_s} = \sum_{i=1}^{n} \frac{\alpha_i}{AF_i} \tag{4.14}$$

where $\lambda_{s\text{-test}}$ is the system failure rate in test condition, $\lambda_{i\text{-test}}$ is the $i$th mechanism failure rate in test condition, $AF_s$ is the system acceleration factor, $AF_i$ is the $i$th mechanism acceleration factor, and $\alpha_i$ is the weight of the $i$th mechanism failure rate in the system failure rate.

A failure rate matrix, Figure 4.1, would be an appropriate form to show the possibility of getting optimum test results for multiple failure mechanism assumption; $\lambda_{p1}, \lambda_{p2}, \ldots, \lambda_{pq}$ are failure rates with fixed high voltage but a variety of temperatures, while $\lambda_{1q}, \lambda_{2q}, \ldots, \lambda_{(p-1)q}$ are those ones related to fixed high temperature but different voltages.

### 4.3.1 TDDB, EM, and HCI Failure Rate Calculations under Single Failure Mechanism Assumption

In a case study, three mechanisms, TDDB, HCI, and EM, are selected to model the Accelerated Life-Test (ALT) test for 90 nm node.

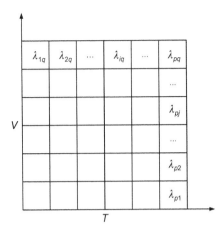

*Figure 4.1 Failure rates from the ALT.*

For TDDB, gate oxide lifetime dramatically shortens with the increase in direct tunneling current, and the dependence of lifetime $t_f$ on gate voltage $V_g$ is given by

$$t_f = A_1 V_g^c \exp\left(\frac{E_{a1}}{KT}\right) \tag{4.15}$$

where $A_1$ is the technology constant. Then

$$\lambda_{TDDB} = A_1 V_g^c \exp\left(\frac{-E_{a1}}{KT}\right) \tag{4.16}$$

The typical range of $E_{a1}$ is from 0.6 to 0.9 eV, and the range of $c$ is from 38 to 45. In this case (90 nm node), the use condition is assumed to be 1.2 volt (voltage) and 20°C (temperature). Furthermore, it is assumed that $E_{a1} = 0.65$ eV and $c = 40$, based on the failure rate being 0.5 FIT at the use condition (for TDDB), which leads to $A_1 = 51.1E + 7$.

Similarly, the failure rate of HCI can be modeled as

$$\lambda_{HCI} = A_2 \exp\left(\frac{-\gamma}{V_d}\right) \exp\left(\frac{-E_{a2}}{KT}\right) \tag{4.17}$$

$E_{a2}$ is reported to be in the range of $-0.1$ to $-0.2$ eV, and the typical range of value of $\gamma$ is reported as 30–80. Assuming that $E_{a2} = -0.15$ eV and $\gamma_2 = 46$ leads to $A_2 = 3.51E + 15$.

The model for EM could be given by

$$\lambda_{EM} = A_3' J^n \exp\left(-\frac{E_{a2}}{KT}\right) \tag{4.18}$$

Generally, $J$ can be estimated by

$$J = \frac{C_{int} \times V}{W \times H} \times f \times \gamma \tag{4.19}$$

where $C_{int}$ is the interconnect capacitance of a specific node, $V$ is the voltage drop across the interconnect segment, $W$ is the interconnect width, $H$ is the interconnect thickness, $f$ is the current switching frequency, and $\gamma$ is the probability that the line switches in one clock cycle. Then, it could be possible to derive

$$\lambda_{EM} = A'_3 \left( \frac{C_{int} \times V}{W \times H} \times f \times \gamma \right)^n \exp\left( -\frac{E_{a3}}{KT} \right) = A_3 V^n \exp\left( -\frac{E_{a3}}{KT} \right) \tag{4.20}$$

The typical value of $n = 2$ and the range of $E_{a3}$ is 0.9–1.2 eV. Assuming that $n = 2$ and $E_{a3} = 0.9$ eV, and the failure rate due to EM is 20 FITs under use condition, then $A_3 = 4.21E + 16$.

Considering independent failure mechanisms, the failure rate at 1.7 V and 140°C would be

$$\lambda_{1.7V,140°C} = \lambda^{TDDB}_{1.7V,140°C} + \lambda^{HCI}_{1.7V,140°C} + \lambda^{EM}_{1.7V,140°C} = 8.93E + 7 \tag{4.21}$$

Similarly, the failure rates in the other tested conditions are shown in Figure 4.2, which can be obtained by the manufacturers in the ALT.

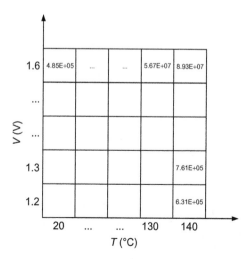

*Figure 4.2 The values of failure rates obtained in the case study.*

The traditional method used by the manufacturers considers a dominant failure mechanism for a system, and therefore the dominant failure mechanism model has been used to extrapolate the failure rate from test conditions:

$$\lambda = A_4 V_g^c \exp\left(-\frac{E_a}{KT}\right) \tag{4.22}$$

where $A_4$, $c$, and $E_a$ are estimated from matrix 7.

In order to find $c$, one could write

$$\lambda = A_4 V_g^c \exp\left(-\frac{E_a}{KT}\right) \tag{4.23}$$

Then by getting logarithm from both sides:

$$\ln \lambda = c \ln V_g - \frac{E_a}{KT} + \ln A_4 \tag{4.24}$$

When the temperature is fixed, there is a linear relation between $\ln \lambda$ and $V_g$, and $c$ is the slope. In practice, due to the long test time, the manufacturers often start the test from relatively high voltage instead of the use voltage. For instance, the voltage test condition in 40°C starts from 1.52 V instead of 1.2 V (as given in Table 4.2). The voltage factor $c$ is dominated by TDDB and is estimated as 38.597, close to 40, which is related to that of TDDB (Figure 4.3).

In order to estimate $E_a$, one can write

$$\ln \lambda = -\frac{E_a}{K} \cdot \frac{1}{T} + \ln A_4 + c \ln V_g \tag{4.25}$$

When the voltage is fixed, there is a linear relation between $\ln \lambda$ and $1/T$, and $-E_a/K$ is the slope. Similarly, in the case of fixed voltage, the temperature test starts from relatively high temperature instead of use temperature. In 1.6 volts, temperature test conditions start from 120°C instead of 20°C. The test data are given in Table 4.3 (only sum can be obtained by the manufacturers in the test).

| Table 4.2 The Traditional Test Conditions under High Voltages and High Temperature | | | | | |
|---|---|---|---|---|---|
| | 1.52 V, 140°C | 1.54 V, 140°C | 1.56 V, 140°C | 1.58 V, 140°C | 1.6 V, 140°C |
| System failure rate | 1.24E + 07 | 2.02E + 07 | 3.31E + 07 | 5.44E + 7 | 8.93E + 07 |

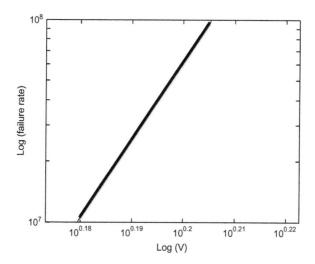

*Figure 4.3 Traditional method for voltage factor estimation.*

| Table 4.3 The Traditional Test Conditions under High Voltage and Low Temperatures | | | | | |
|---|---|---|---|---|---|
| | 1.6 V, 120°C | 1.6 V, 125°C | 1.6 V, 130°C | 1.6 V, 135°C | 1.6 V, 140°C |
| System failure rate | 3.52E + 07 | 4.48E + 07 | 5.67E + 07 | 7.14E + 07 | 8.9E + 07 |

Then, $-E_a/K$ can be estimated as $-7558.8$ and $E_a$ is equal to 0.651.

Since it is supposed that TDDB is the dominant failure mechanism, the estimation of $E_a$ is dominated by TDDB and $E_a$ is approximately equal to that of TDDB (Figure 4.4).

Then the failure rate under use condition (1.2 V and 20°C) can be extrapolated from the data provided under the test condition (1.6 V and 140°C). The system acceleration factor can be calculated as

$$AF_s = AF_t \cdot AF_V = \frac{A_4 V_{g\text{-test}}^c \exp(-E_a/KT_{test})}{A_4 V_{g\text{-use}}^c \exp(-E_a/KT_{use})} = 1.02E + 8 \quad (4.26)$$

Therefore

$$\lambda_{1.2V,20°C} = \frac{\lambda_{1.7V,140°C}}{AF_s} = \frac{8.93E + 7}{1.20E + 8} = 0.75 \quad (4.27)$$

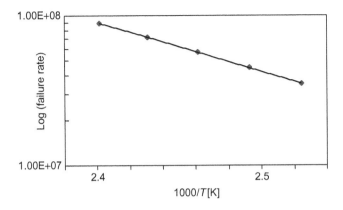

*Figure 4.4 Estimation of* $E_a$.

The failure rate estimated from the test data (and provided by manufacturers) is much smaller than that calculated from the real failure rate under use condition (50.5 FITs). In this case, since TDDB is supposed to be the mere dominant failure mechanism, the related AF and failure rate are calculated under high-voltage conditions and those of others are completely ignored. Therefore, the failure rate extrapolated from high temperature and voltage is essentially that of TDDB under the use condition, which is only a piece of the puzzle. The whole picture would appear when data comes from high-voltage/high temperature, high-voltage/low temperature, and high temperature/low-voltage test conditions.

### 4.3.2 TDDB, EM, and HCI Failure Rate Calculations under Multiple Failure Mechanism Assumption

To show the difference, the previous case study is recalculated with the multiple failure mechanism approach together with the suggested test matrix; the model and parameters of each mechanism are the same as those in the traditional approach. All the failure rates of TDDB, HCI, and EM under use condition are also assumed to be 20 FITs.

As mentioned above, TDDB dominates under fixed high temperature with increasing high voltage, and the data in this field can be regarded as the TDDB failure rate approximately. Its estimation process, including the data, is the same as that in previous sections. Then, TDDB failure rate would be $\lambda_{1.2V,\,20^\circ C}^{TDDB} = 0.75$ FITs.

For HCI, using the previous equation:

$$\lambda_{\text{HCI}} = A_2 \exp\left(-\frac{\gamma_2}{V_d}\right)\exp\left(-\frac{E_{a2}}{KT}\right) \qquad (4.28)$$

then

$$\ln \lambda_{\text{HCI}} = -\frac{\gamma_2}{V_d} - \frac{E_{a2}}{KT} + \ln A_2 \qquad (4.29)$$

When the temperature is fixed, there is linear relation between $\ln \lambda_{\text{HCI}}$ and $1/V_d$, and $-\gamma_2$ is the slope. HCI dominates under fixed high voltage and low temperature; therefore, the related data could be regarded as the HCI failure data as shown in Figure 4.5, i.e., the voltage factor is $\gamma_2 = 47.313$ (Table 4.4).

Since the acceleration test data has the same temperature as the one under use condition, only voltage acceleration factor is needed to be calculated for HCI acceleration factor:

$$\text{AF}_{\text{HCI}} = \frac{A_2 \exp\left(-\gamma/V_{d\text{-test}}\right)\exp\left(-E_{a2}/KT_{\text{test}}\right)}{A_2 \exp\left(-\gamma/V_{d\text{-use}}\right)\exp\left(-E_{a2}/KT_{\text{use}}\right)} = 1.91\text{E} + 4 \qquad (4.30)$$

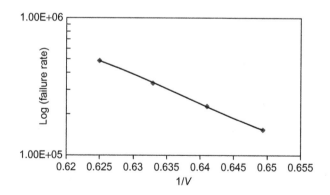

*Figure 4.5 Estimation of voltage acceleration factor $\gamma_2$ (HCI).*

| Table 4.4 Failure Rate with Fixed Low Temperature and High Voltages (HCI) | | | | | |
|---|---|---|---|---|---|
| | 1.52 V, 20°C | 1.54 V, 20°C | 1.56 V, 20°C | 1.58 V, 20°C | 1.6 V, 20°C |
| Failure rate | 7.03E + 05 | 1.53E + 05 | 2.27E + 05 | 3.33E + 06 | 4.85E + 06 |

and

$$\lambda_{1.2V,20°C} = \frac{\lambda_{1.6V,20°C}^{HCI}}{AF_{HCI}} = 25.4 \text{ FITs} \qquad (4.31)$$

To estimate the activation energy of EM, one can use

$$\lambda_{EM} = A_3 V^n \exp\left(-\frac{E_{a3}}{KT}\right) \qquad (4.32)$$

Then by getting logarithm from both sides

$$\ln \lambda_{EM} = -\frac{E_{a3}}{KT} + n \ln V + \ln A_3 \qquad (4.33)$$

When the voltage is fixed, there is a linear relation between $\ln \lambda_{EM}$ and $1/T$, and $-E_{a3}/K$ is the slope. Since EM dominates under fixed high temperature and low voltages, the related could be regarded as the EM failure rate approximately. Then the voltage factor would be $E_{a3}/K = -10,439$, which leads to $E_{a3} = 0.9$ (Table 4.5).

However, since the acceleration test data has the same voltage as the data under use condition, only $E_{a3}$ is needed to be calculated for EM acceleration factor:

$$AF_{EM} = \frac{A_3 V_{test}^n \exp\left(-E_{a3}/KT_{test}\right)}{A_3 V_{use}^n \exp\left(-E_{a3}/KT_{use}\right)} = 3.15E + 4 \qquad (4.34)$$

which leads to

$$\lambda_{1.2V,20°C} = \frac{\lambda_{1.2V,140°C}^{EM}}{AF_{EM}} = 20.0 \text{ FITs} \qquad (4.35)$$

Therefore, the system failure rate could be calculated from the following equation:

$$\lambda_{1.2V,20°C}^{s} = \lambda_{1.2V,20°C}^{TDDB} + \lambda_{1.2V,20°C}^{HCI} + \lambda_{1.2V,20°C}^{EM} = 46.15 \text{ FITs} \qquad (4.36)$$

The results of this method are by far closer to those of use conditions (50.5 FITs), which shows that it is more accurate than the one given by traditional single failure mechanism approach together with the related test data.

| Table 4.5 Failure Rate with Fixed Low Temperature and High Voltages | | | | | |
|---|---|---|---|---|---|
| | 1.2 V, 120°C | 1.2 V, 125°C | 1.2 V, 130°C | 1.2 V, 135°C | 1.2 V, 140°C |
| Failure rate | 1.74E + 05 | 2.43E + 05 | 8.41E + 05 | 3.371E + 05 | 6.31E + 05 |

## 4.4 ELECTRONIC SYSTEM CFR APPROXIMATION/ JUSTIFICATION

The Achilles' heel of Mil-handbook 217F is the assumption of CFR for the new generation of electronic devices. It was said that the assumption of CFR could only be valid in the days of vacuum tubes not for solid-state components; the failure rate of solid-state components not only decreases but also approaches zero. Moreover, the use of CFR gives a pessimistic result for the new devices.

Recent studies based on the physics of failure (PoF) not only revealed different failure mechanisms for microelectronic devices and circuits but also modeled their failure distributions. It was believed that almost all the failure mechanisms were in the wear-out part of bathtub curve. However, as technology progressed, the new era of "constant rate processes" approximation is going to dominate the reliability of microelectronic device and circuit. The "constant rate processes" assumption is the result of failure mechanisms shifting from wear-out part to CFR part due to deep submicron technologies as well as the shrinkage of transistor geometries. TDDB Weibull slope approach to one is one of the significant sign of the shift.

However, debate over the last decade of the disadvantages and uselessness of Mil-handbook 217F makes the "constant rate processes" assumption almost unusable. On one hand, experts don't want to build the new handbook on the weak bricks of Mil-handbook 217F; on the other hand, the memorylessness property of CFR is another obstacle in the way of accepting the new approach.

Since, in the absence of a dominant failure mechanism in microelectronic chip/system reliability, all failure distributions could be mixed with the same wear-out drivers as voltage, temperature, and frequency, it is useful to review the result of distribution mixtures and their properties. The counterintuitive result and the interpretations of distribution mixture properties would help to digest the "constant rate process" approach and its memorylessness characterization. The decreasing failure rate (DFR) of a bunch of increasing failure distribution mixtures together with the "negative ageing" property give a new kind of perspective to the reliability issues.

Drenick's theorem is another way of justifying the CFR approximation for a system consisting of a very large number of components like a microelectronic chip.

In microelectronic system reliability, one may define a system as an integrated collection of subsystems, modules, or components, which function together to attain specific results. A system fails because some or all of its components do. It is obvious that there is a connection between the reliability of components of a system and that of the system as a whole. The system reliability is given as a function of the reliability of its components. It is not easy or simple to formulate this functional dependence. Therefore, it is common to simplify the case as the system with independent components time to failure, and simple coherent structure functions, like those of series, parallel, and cross-linked. In practice, however, exact evaluation of system reliability is extremely difficult and sometimes impossible. Once one obtains the expression for the structure function, the system reliability computations become straightforward [76].

This section tries to claim that the "CFR" approach could be a good approximation to predict the microelectronic system reliability. However, like any other statistical one, it needs to be justified by gathering more field data; comparing the results of simulation with the actual results would be the best justification for the new approach.

## 4.4.1 Exponential Distribution

The exponential distribution is one of the key distributions in the theory and practice of statistics. It possesses several significant properties—most notably, its characterization through the lack of memory property. Yet it exhibits great mathematical tractability. Consequently, there is a vast body of literature on the theory and applications of the exponential distribution.

Exponential distributions are encountered as lifetime distributions with constant hazard rate. If the cumulative distribution function of $X$, a random variable representing lifetime, is $F_X(x)$ and the constant hazard rate is $\lambda$, then

$$-\frac{\mathrm{d}\log(1 - F_X(x))}{\mathrm{d}x} = \lambda \quad (x > 0) \tag{4.37}$$

or, in terms of the survival function $\overline{F}_X(x) = 1 - F_X(x)$,

$$-\frac{d \log \overline{F}_X(x)}{dx} = \lambda \quad (x > 0) \tag{4.38}$$

Solving the differential equation by applying $\overline{F}_X(0) = 1$, then

$$\overline{F}_X(x) = e^{-\lambda x} \quad (x > 0) \tag{4.39}$$

and the corresponding probability density function of $X$ as

$$f_X(x) = \lambda e^{-\lambda x} \quad (x > 0) \tag{4.40}$$

If $\lambda = 1$, then the standard exponential distribution with the survival function:

$$\overline{F}_X(x) = e^{-x} \quad (x > 0) \tag{4.41}$$

and the probability density function:

$$f_X(x) = e^{-x} \tag{4.42}$$

For $y > 0$ and $x > 0$:

$$P[X > x + y \mid X > x] = P[X > y] \tag{4.43}$$

i.e., the future lifetime distribution, given survival to age $x$, does not depend on $x$. This property, termed lack of memory or, equivalently, lack of ageing, characterizes exponential distributions, as can easily be seen. The last equation can also be written equivalently as

$$\frac{S_X(x + y)}{S_X(x)} = S_X(y) \tag{4.44}$$

i.e.,

$$\log S_X(x + y) = \log S_X(x) + \log S_X(y) \tag{4.45}$$

Differentiating the above equation with respect to $y$:

$$-\frac{d \log S_X(x + y)}{dy} = -\frac{d \log S_X(y)}{dy} \quad [= \lambda_X(y)] \tag{4.46}$$

since

$$-\frac{d \log S_X(x + y)}{dy} = -\frac{d \log S_X(x + y)}{d(x + y)} \quad [= \lambda_X(x + y)] \tag{4.47}$$

The hazard rate $y$ is the same as $(x + y)$ for any $x$ and $y$, i.e., it is constant.

This lack of memory property is a critical part of many of the statistical analyses based on the exponential distribution. Exponential distributions are commonly employed in the formation of models of lifetime distributions and stochastic processes in general. Even when simple mathematical expressions of the distribution are inadequate to describe real-life complexity, exponential (Poisson) often serves as a benchmark with reference to which affects of departures to allow for specific types of disturbance can be assessed [75].

The exponential distribution is remarkably friendly. Closed form expressions exist for its density, distribution, moment generation function, mean residual life function, failure rate function, moments of order statistics, record values, etc. So, everyone introducing a new concept or functional to classify or organize distributions inevitably uses the exponential distribution as one of their examples. The exponential distribution is often the simplest example or the most analytically tractable. A consequence of its tendency to represent the extreme case is numerous characterizations.

The most important characterizations of exponential distribution could be classified as

- Lack of memory results
- Distributional relations among order statistics
- Independence of functions of order statistics
- Moments of order statistics
- Geometric compounding
- Record value concepts

Mixtures of distributions of lifetimes occur in many settings. In engineering applications, it is often the case that populations are heterogeneous, often with a small number of subpopulations. The concept of a failure rate in these settings becomes a complicated topic, especially when one attempts to interpret the shape as a function of time. Even if the failure rates of the subpopulations of the mixture have simple geometric or parametric forms, the shape of the mixture is often not transparent. Recent results focus on general results to study whether it is possible to provide approximation for a system.

## 4.4.2 The Reliability of Complex Systems

Mixtures are a common topic in most areas of statistics. They also play a central role in reliability and survival analysis. However, the failure rate of mixed distributions is a source of much confusion.

The density function of a mixture from two subpopulations with density functions $f_1$ and $f_2$ is simply given by

$$F(t) = pf_1(t) + (1-p)f_2(t), \quad t \geq 0, \ 0 \leq p \leq 1 \tag{4.48}$$

and thus the survival function of the mixture is also a mixture of the two survival functions, i.e.,

$$\overline{F}(t) = p\overline{F}_1(t) + (1-p)\overline{F}_2(t) \tag{4.49}$$

The mixture failure rate $r(t)$ obtained from failure rates $r_1(t)$ and $r_2(t)$ associated with $f_1$ and $f_2$, respectively, can be expressed as

$$r(t) = \frac{pf_1(t) + (1-p)f_2(t)}{p\overline{F}_1(t) + (1-p)\overline{F}_2(t)} \tag{4.50}$$

where $f_i(t)$ and $\overline{F}_i(t)$ are the probability density and survival functions of the distribution having failure rate $r_i(t)$, $i = 1, 2$.

Let us assume for $r(t)$, then

$$r(t) = h(t)r_1(t) + (1 - h(t))r_2(t) \tag{4.51}$$

where

$$h(t) = \frac{1}{1 + g(t)}, \quad g(t) = \frac{(1-p)\overline{F}_2(t)}{p\overline{F}_1(t)}$$

Clearly $0 \leq h(t) \leq 1$.

The above equation can be easily generalized to accommodate mixtures of $k$ subpopulations giving

$$r(t) = \frac{\sum_{i=1}^{k} p_i f_i(t)}{\sum_{i=1}^{k} p_i \overline{F}_i(t)} \tag{4.52}$$

where

$$i = 1, 2, 3, \ldots, k, \ 0 < p_i < 1, \ \sum_{i=1}^{k} p_i = 1, \ k \geq 2.$$

It has been proved that a mixture of two DFR distributions is again DFR.

As mentioned before, the first kind of counterintuitive case is that studied by Proschan [78]. Proschan worked on the pooled data for airplane air conditioning systems whose lifetimes are known to be exponential that exhibit a DFR. Because DFRs are usually associated with systems that improve with age, this was initially thought to be counterintuitive [75].

Pooled data on the time of successive failures of the air conditioning systems of a fleet of jet airplanes seemed to indicate that the life distribution had the DFR. More refined analysis showed that the failure distribution for each airplane separately was exponential but with a different failure rate. We can assume here the theorem that a mixture of distributions having non-IFR (a mixture of exponential distributions, for instance) also has a non-IFR, the apparent DFR of the pooled air conditioning life distribution was satisfactorily explained. This has implications in other areas, where an observed DFR may well be the result of mixing exponential distributions having different parameters.

Proschan brings two hypotheses called $H_0$ and $H_1$ as follows: if all the planes under investigation had the same failure rate, then the failure intervals pooled together for the different planes would be governed by a single exponential distribution ($H_0$ hypothesis); on the other hand, if each plane corresponded to a different failure rate, it would then follow that distribution of the pooled failure intervals would have a DFR ($H_1$ hypothesis). To validate this inference, he invokes the following theorems from Barlow, Marshall, and Proschan [78].

The theorem says: if $F_i(T)$ has a DFR, $i = 1, 2, 3, \ldots, n$, then

$$G(t) = \sum_{i=1}^{n} p_i F_i(t) \tag{4.53}$$

has a DFR, where each

$$p_i > 0, \quad \sum_{i=1}^{n} p_i = 1 \tag{4.54}$$

After proving the theorem, he rejects $H_0$ in favor of $H_1$ and concludes that the pooled distribution has a DFR, as would be expected if the individual airplanes each displayed a different CFR.

Another anomaly, at least to some, was that mixtures of lifetimes with IFRs could be decreasing on certain intervals. A variant of the above is due to Gurland and Sethuraman [77], which gives examples of mixtures of very rapidly IFRs that are eventually decreasing.

We recall here that the mixture of exponential distributions (which have CFR) will always have the DFR property (shown by Proschan [78]) and referring to a converse result [79] who demonstrates that any Gamma distribution with shape parameter less than 1 and therefore a DFR distribution can be expressed as a mixture of exponential distributions, they tried to show that it is reasonable to expect that mixture of IFRs distributions that have only moderately IFRs can possess the DFR property and gave some examples as well. A detailed study of classes of IFR distributions, whose mixtures reverse the IFR property, was given in 1993. They also studied the class of IFR distributions that when mixed with an exponential becomes DFR. A few examples of distributions from this class are the Weibull, truncated extreme, Gamma, truncated normal, and truncated logistic distributions [77].

By considering that the mixtures of DFR distributions are always DFR, and some mixtures of IFR distributions can also be ultimately DFR, Gurland and Sethuraman studied various types of discrete and continuous mixtures of IFR distributions and developed conditions for such mixtures to be ultimately DFR [77].

They showed the unexpected results that the mixture of some IFR distributions, even those with very rapidly IFRs (such as Weibull, truncated extreme), become ultimately DFR distribution. In practice, data from different IFR distributions are sometimes pooled, to enlarge sample size, for instance. These results serve as a warning note that such pooling may actually reverse the IFR property of the individual samples to an ultimately DFR property for the mixture. This phenomenon is somewhat reminiscent of Simpson's paradox, wherein a positive partial association between two variables may exist at each level of third variable, yet a negative overall unconditional association holds between the two original variables.

They provide the condition on mixture of two IFRs to show that the result has DFR, while the mixture of IFR distribution functions is given by

$$p(t) \equiv p_1 F_1(t) + p_2 F_2(t) \tag{4.55}$$

where

$$0 \le p_1, \ p_2 \le 1$$

and

$$p_1 + p_2 = 1$$

. The provided conditions lead to interesting results that certain mixtures of IFR distributions, even those with very rapidly IFRs, become DFR distributions [77].

When considering the reliability of systems, an IFR seems to be very reasonably as age, use or both may cause the system to wear out over time. The reasons of DFR, i.e., a system improves as time goes by, are less intuitive. Mixtures of lifetime distributions turn out to be the most widespread explanation for this "positive ageing." The more paradoxical case corresponds to the exponential distribution that shows a CFR while exponential mixtures belong to the DFR class. Proschan found that a mixture of exponential distributions was the appropriate choice to model the failures in the air conditioning systems of planes. Such mixing was the reason of the DFR that the aggregated data exhibited. The DFR and the decreasing failure rate average (DFRA) classes are closed under mixtures. A similar result does not hold for the IFR class; however, many articles focus on mixtures of IFR distributions that reverse this property over time and exhibit a DFR as time elapses.

By using a Cox's proportional hazard rate model, a nonnegative random variable and the conditional failure rate are defined; the conditional failure rate is in the form of a Gamma distribution applied to an increasing Weibull distribution. The result shows that the greater the mean of mixing distribution, the sooner the increasing failure rate average (IFRA) property is reversed.

In the survival analysis literature, it is known that if an important random covariate in a Cox model is omitted, the shape of the hazard rate is drastically changed. Other types of articles mention that in many biological populations, including humans, lifetime of organisms at extreme old age exhibits decreasing hazard rate. A natural question to ask is whether this means that some of the individuals in the population are improving or not, which leads to misconception and misunderstanding.

In most settings involving lifetimes, the population of lifetimes is not homogeneous. That is, all of the items in the population do not have exactly the same distribution; usually a percentage of the lifetimes

are of a type different from the majority. For example, in most industrial populations there is often a subpopulation of defective items. For
electronic components, most of the population might be exponential,
with long lives, while a small percentage often has an exponential distribution with short lives. Even though each of the subpopulations has
CFR, the mixed population does not. Proschan [78] observed that such
a population has DFR. An intuitive explanation is that there are stronger (i.e., lower failure rate) and weaker (i.e., higher failure rate) components and as time goes by the effect of the weaker component
dissipates and the stronger takes over. Another way of saying this is
that the failure rate of the mixture approaches the stronger failure rate.
Although this had been observed in various special cases, one of the first
general results appears in [79]. Gurland and Sethuraman [77] observed
that mixtures of components which had fairly rapidly IFRs could still
turn out to have eventually DFRs when mixed. It is important to know
the behavior which occurs when populations are pooled. This pooling
can occur naturally or can be done by statisticians to increase sample
size. Besides, for modeling purposes it is often useful to have available a
distribution which has a particular failure shape. For example, it is useful to know how to pool distributions to obtain the important bathtub-
shaped failure rate.

Moreover, studying the tail behavior of the failure rate of mixtures
of lifetime distributions, one could conclude that if the failure rate of
the strongest component of the mixture decreases to a limit, then the
failure rate of the mixture decreases to the same limit. For a class of
distributions containing the gamma distributions this result can be
improved in the sense that the behavior of the failure rate of the mixture asymptotically mirrors that of the strongest component in whether
it decreases or increases to a limit [79].

Some general rules of thumb concerning the asymptotic behavior of
the failure rate when several distributions are mixed were already found.
First, the failure rate of the mixture will approach the stronger (lowest)
failure rate so that there is a downward trend. However, if the strongest
failure rate is eventually increasing, the mixture will become increasing. If
one of the mixture probabilities is close to one, the mixture failure rate
will initially behave like that component. If the component with probability close to one becomes the strongest component, then the mixture will
eventually behave like that component. If the failure rates cross, then the

point of intersection is also a factor. Differences of the $y$-intercepts and the ratio of the slopes also play a role [80].

The study of the life of human beings, organisms, structures, materials, etc. is of great importance in the actuarial, biological, engineering, and medical sciences. It is clear that research on ageing properties is currently pursued. While positive ageing concepts are well understood, negative ageing concepts (life improved by age) are less intuitive. As mentioned before, there have been cases reported by several authors where the failure rate functions decrease with time. Sample examples are the business mortality, failures in the air conditioning equipment of a fleet of Boeing 720 aircrafts or in semiconductors from various lots combined, and the life of integrated circuits modules. In general, a population is expected to exhibit DFR when its behavior over time is characterized by "work hardening" in engineering terms or "immunity" in biological terms. Modern phenomenon of DFR includes reliability growth in software reliability.

### 4.4.3 Drenick's Theorem

Drenick published a paper in which he proved that, under certain constraints, systems which are composed of a "large" quantity of nonexponentially distributed subcomponents tend themselves toward being exponentially distributed. This profound proof allows reliability practitioners to disregard the failure distributions of the pieces of the system since it is known that the overall system will fail exponentially. Given that most systems are composed of a large number of subcomponents, it would seem that Drenick's theorem is a reliability analysis godsend. The usefulness of this theorem, however, lays in the applicability of the proof constraints.

Kececioglu delineated the constraints of Drenick's theorem quite well as follows:

1. The subcomponents are in series.
2. The subcomponents fail independently.
3. A failed subcomponent is replaced immediately.
4. Identical replacement subcomponents are used.

If the four conditions above are met, then as the number of subcomponents and the time of operation tend toward infinity, system failures tend towards being exponentially distributed regardless of the nature of the subcomponents' failure distributions [81].

In his paper, Drenick emphasizes the role of many-component systems and says that "In theoretical studies of equipment reliability, one is often concerned with systems consisting of many components, each subject to an individual pattern of malfunction and replacement, and all parts together making up the failure pattern of the equipment as a whole." His work is concerned with that overall pattern and more particularly with the fact that it grows the more simple, statically speaking, the more complex the equipment. Under some reasonably general conditions, the distribution of time between failure tends to the exponential as the complexity and the time of the operation increase; and somewhat less generally, so does the time up to the first failure of the equipment. The problem is in the nature of the probabilistic limit theorem and that accordingly the addition of independent variables and the central limit theorem is a useful prototype.

It may be useful to add several comments, some concerned with the practice of reliability work and others with the theory.

1. It has been assumed that the failure incidents in one component of a piece of equipment are statistically independently of the rest. The independence among failures is more likely to be satisfied in well-designed than in poorly designed equipment.
2. It always makes good sense to lump the failures of many, presumably dissimilar, devices into one collective pattern.
3. In the long run, the failure pattern of a complex piece of equipment is determined essentially by the mean lives of its components.
4. There is a distinction between (initial) survival probability and the residual one; although both are exponential, they have different means. Therefore, despite the similarity of two distributions, one should not use them interchangeably. Otherwise, it gives a pessimistic kind of result, for instance.
5. Mathematically, even the weak form of limit law condition is sufficient for the asymptotic convergence of one distribution to the exponential one.

Drenick's assumption would be considered a mathematical justification for the microelectronic system CFR approach [81,82].

## 4.5 PoF-BASED CIRCUITS RELIABILITY PREDICTION METHODOLOGY

With a history of more than five decades, reliability prediction has an important role in business decisions such as the system design, parts selection, qualification warranties, and maintenance. Nowadays electronic system designers have their own industry-specific reliability prediction tools, such as the well-known MlLHDBK-217, SAE reliability prediction method, Telcordia SR-332, and PRISM [83,84]. Many of those methods are empirically based, which were built upon field data and extrapolations such as the parts count method and the parts stress methods with various kinds of prefactors [85]. One big disadvantage of those empirical-based methods is the lack of integration of the PoF models because of the complexity and difficulty for the system designers to get detailed technology and microcircuit data. Prediction accuracy is diminished without those PoF models, and the situation is becoming worse with technology advancement. Today's microelectronic devices, featuring ultrathin gate oxide and very short channels, suffer various detrimental failure mechanisms, including TDDB [86], NBTI (negative bias temperature instability) [87], HCI [88], and EM, as the nonideal voltage scaling brings higher field and current density. As mentioned before, each failure mechanism has its unique dependence on voltage and temperature stresses and all can cause device failure. The traditional prediction method is not applicable as before, considering the multiple failure mechanisms' effect and the difficulty to obtain enough up-to-date field data. Device manufacturers face the same challenge to maintain and further improve reliability performance of advanced microelectronic devices, in spite of all kinds of difficulties from technology development, system design, and mass production. Conventional product reliability assurance methods, such as burn-in HTOL and HALT, are gradually losing competitiveness in cost and time because the gap between normal operating and acceleration test conditions is continuing to narrow and the increased device complexity makes sufficient fault coverage tests more expensive. An accurate reliability simulation and prediction tool is greatly needed to guide the manufacturers to design and deploy efficient qualification procedure according to the customer's need, and help the designers to get in time reliability feedback to improve the design and guarantee the reliability at the very first stage.

The needs of accurate reliability prediction from both device and system manufacturers require integration of PoF models and statistical models into a comprehensive product reliability prediction tool that takes the device and application details into account. A new PoF-based statistical reliability prediction methodology is proposed in this chapter. The new methodology considers the needs of both the device manufacturer and the system supplier by taking application and design into account. Based on circuit level operation-oriented PoF analysis, this methodology provides an application-specific reliability prediction, which can be used to guide qualification and system design.

The prediction methods based on the handbook approach usually provide conservative failure rate estimation [89]. Many things can and should be done to improve the prediction accuracy. The two important considerations are as follows:

1. Integration of PoF analysis and modeling. This is a prominent issue because advanced microelectronic devices are vulnerable to multiple failure mechanisms. These failure mechanisms have their unique voltage and temperature dependence. No unified lifetime model can take all these into account.
2. Integration of failure mechanism lifetime distribution. CFR assumption might give fast and cost-effective failure rate estimation at the system level because the inherent inaccuracy is so obvious without justification from detailed PoF analysis. Weibull distribution has been demonstrated as the best-fit lifetime distribution for TDDB. Lognormal distribution has been experimentally verified for HCI and EM. These specific lifetime distributions should be utilized in prediction for better modeling component and system lifetime.

### 4.5.1 Methodology

The Physics of Failure (PoF)-based statistical approach models device reliability by considering all the intrinsic failure mechanisms under dynamic stresses. Generally speaking, today's microelectronic device integrates many functional blocks, which consist of thousands or even millions of transistors. Running a full spectrum simulation will consume unacceptable amounts of resources. To reduce the simulation complexity and release the heavy load of computation, the proposed PoF statistical reliability prediction methodology takes four unique approaches by considering the repetitive characteristic of CMOS circuits:

1. Cell-based reliability characterization. Standard cells (inverter, NOR, NAND, etc.) are the fundamental building blocks in modern

VLSI circuit design. Cells in the same category have similar structures and operation profiles. For instance, the SRAM chip consists of millions of hit cells that have the same transistor configuration. These SRAM bit cells have similar operations applied to them: read, write, or hold. For chips as complex as a microprocessor, it can still be divided into functional block and further the cells. In the PoF statistical approach, reliability characterization starts from the standard cells. Doing this can ease system designers' concern of understanding circuit details and running circuit simulation. Advantages of cell-based reliability characterization are listed below:

a. Time saving in circuit design. Cell schematic and layout can be obtained from design kits provided by semiconductor manufacturers or design houses. There is no need for system designers to understand the circuits from the very beginning since they often are not electrical engineers. What they need to understand is the categorization of the cells and the reliability character of each category.

b. Time saving in circuit simulation. The VLSI device simulation requires detailed circuit information and consumes lots of computation resource. Since the goal of circuit simulation is to find out the stress profile of transistors, cell level simulation provides a better option because of the small count of transistors inside a cell.

2. Equivalent stress factor. Equivalent stress factor (ESF) is used to convert dynamic stresses to static stresses that have the same degradation effect. For each failure mechanism, lifetime model is built upon acceleration tests, which are generally carried out with highly acceleration static voltage and temperature stresses. However, a transistor in real operation has a dynamic stress profile and the static PoF models can't be applied directly. The ESFs are obtained through cell reliability characterization and then applied in device reliability prediction. These factors are specified to cell, transistor, operation, and failure mechanism. Given a specific cell, for each operation, the voltage and current stresses of each transistor in the cell are obtained through SPICE simulation. For each failure mechanism, degradation under different stress conditions is accumulated and converted to an equivalent total stress time (ETST) under a specified static stress condition by utilizing appropriate acceleration models. The ESF is the ratio of the ETST to the real stress time. To estimate cell reliability in a real application, the cell operation

profile needs to be determined first. The next step is using ESFs to calculate the "effective" stress time for each failure mechanism of each transistor. The cell reliability is estimated as a series system in which each transistor inside the cell corresponds to a component.

3. Best-fit lifetime distribution. To improve the prediction accuracy, the best-fit lifetime distribution for each failure mechanism should be taken instead of using the CFR model without justification. In the PoF statistical approach, Weibull distribution is used to model TDDB failures, and lognormal distribution for NBTI, HCI, and EM. The cell is considered as a series system with each component corresponding to one failure mechanism.

4. Time-saving chip-level interconnect EM analysis. Chip-level EM analysis becomes more important as IC complexity is always driven up by scaling. Although EM becomes more serious in submicron designs, it is limited to the power distribution network in most cases [90]. Circuit designers must follow design rules, which set the interconnect dimension and current limits. Many industry tools have been developed to help circuit designers check the EM hotspot, such as VoltageStorm™ from Cadence Design Systems and RailMill™ from Synopsys. To optimize the EM resistance, lower level interconnects are designed to be EM-failure-free by considering the Blech effect [91]. The power network becomes the weakest link because of the large current density it carries and the local Joule heating effect. This has been verified by acceleration test results [92].

In the PoF statistical approach, chip-level EM analysis is focused on the power network since all designs should pass the design rules check, and final products must survive the high temperature, high-voltage defect screening. This provides a good approximation without running full-detailed interconnect network EM analysis.

### 4.5.2 Assumptions

The assumptions of the PoE-based statistical approach are briefly explained below:

1. Degradation discretion and accumulation. For each failure mechanism, the degradation process can be discretized by dividing the whole stress period into small intervals in order to accurately model the dynamic stress.

2. Negligible NBTI recovery effect. NBTI degradation has been observed to have a recovery effect in acceleration tests after the stress has been removed [93,94]. Physical understanding of these phenomena is still not clear. Since NBTI is a long-term reliability concern, and recovery disappears quickly when the stress is reapplied, the worst case NBTI is considered in the PoF statistical approach.

3. Independent failure mechanisms. All the failure mechanisms are assumed to be independent. Each failure mechanism has its specific degradation region inside the transistor. TDDB causes damage inside the gate oxide while HCI/NBTI increase interface trap density. For PMOS, HCI and NBTI have been reported to be independent. There is no confirmation about interaction in field failure from the literature research.

4. Competing failure modes. A device is treated as a series system in which any cell failure will cause device failure. Every cell is viewed as a series system with each failure mode composing a block of the series system.

### 4.5.3 Input Data

In order to carry out reliability simulation and prediction, all the following information needs to be gathered: device application profile, device structure, cell schematic and layout, failure mechanism lifetime model, and statistical distribution.

1. Application profile. The device application profile can be broken down into operation phases with distinguishable environment factors. The PoP statistical approach deals with intrinsic failure mechanisms only, and the input data of each operation phase should include the ambient temperature (TA) and the operating status-power on hours (PoH). Other factors such as humidity and vibration are not considered since they are mostly related to mechanical reliability.

2. Device structure and operation. The PoF statistical approach takes a divide-and-conquer way to reduce the complexity of reliability simulation of VLSI devices. A device functional diagram is needed to divide the whole chip into functional blocks. Inside each functional block, cells are categorized and analyzed. Device operation needs to be analyzed to build cell operation profile.

3. Cell reliability simulation inputs including cell schematic and layout, technology file (this should be obtained from device

manufacturers), SPICE models, and stimuli file (the stimuli file should be application oriented so that the reliability output can be directly correlated to the stresses in real application).

4. Failure mechanism models and parameters.

- Lifetime models and parameters. With the technology information ($t_{ox}$, $V_d$, etc.) and acceleration test data, reliability engineers can choose or build the appropriate lifetime model for each failure mechanism. Once the lifetime models have been decided, the model parameters can be estimated from maximum likelihood estimation (MLE) analysis or other regression analysis of acceleration test data.

- Failure distributions and parameters. It is important to have the correct failure distributions to estimate device reliability. For EM, lognormal distribution is normally the first choice. Weibull distribution has been widely used to model TDDB failures. For HCI and NBTI, lognormal distribution can be utilized. With given acceleration test data, the goodness-of-fit of these statistical distributions can be checked and the related distribution parameters can be estimated.

## 4.5.4 Device Thermal Analysis

Power dissipation of modern microelectronic devices has been rapidly increasing along with increasing transistor counts, clock frequencies, and subthreshold leakage currents. The maximum power consumption of Intel microprocessors has been observed to increase by a factor of a little more than 2× every four years [95]. Detailed device thermal analysis becomes more important since most of the failure mechanisms are thermally activated. TDDB, NBTI, and EM all have their own positive activation energy; only HCI has a negative activation energy, which means HCI degrades faster at lower temperatures. For VLSI devices, thermal analysis should be carried out at functional block level because each block may have its own application pattern and the temperature across the whole chip might not be uniform. Several tools have been developed to do detailed chip-level thermal—electrical analysis, such as ILLIADS [96] and HotSpot [97]. However, running this kind of tool requires very detailed circuit information and takes time for a system engineer to understand and learn. In the PoF statistical approach, device thermal analysis is carried out at the functional block level in order to get quick and reasonable temperature estimation. Each functional block is assumed to have a uniform temperature.

With a given application profile, the average substrate temperature of a block can be estimated as

$$T_S = T_A + P_{total} \cdot R_{sa} \tag{4.56}$$

where $P_{total}$ is the total power dissipation of the block and $R_{sa}$ is the substrate-to-ambient thermal resistance. $R_{sa}$ can be approximated by [96]

$$R_{sa} = \frac{1}{10^5 \times \text{DieSize}} \tag{4.57}$$

$P_{total}$ is comprised of three main sources: dynamic switching power $P_{dyn}$, leakage power $P_{leakage}$, and switching power $P_{sw}$.

$$P_{total} = P_{dyn} + P_{leakage} + P_{sw} \tag{4.58}$$

$P_{dyn}$ is due to charging and discharging of capacitive load, which mainly consists of wiring capacitance [97,98].

$$P_{dyn} = a \cdot C_{total} \cdot V_d^2 \cdot f_c \tag{4.59}$$

where $a$ is the activity coefficient, $V_d$ is the supply voltage, $C_{total}$ is the total capacitive load of the wiring network, which can be obtained through RC extraction, and $f_c$ is the clock frequency.

There are two sources of leakage power: diode leakage power and subthreshold leakage power [96]. The diode leakage power is often neglected because it is generally small when compared with other power components. $P_{leakage}$ can then be expressed by

$$P_{leakage} = V_d \cdot I_{sl} \tag{4.60}$$

where $I_{sl}$ is the subthreshold leakage current and

$$I_{sl} = K_0 \cdot \exp\left(\frac{V_{gs} - V_{th}}{nV_t}\right) \cdot \left(1 - \exp\left(-\frac{V_{ds}}{V_t}\right)\right) \tag{4.61}$$

where $K_0$ is a function of technology, $V_t$ is the thermal voltage ($kT/q$), and $V_{th}$ is the threshold voltage, and

$$n = 1 + \frac{\varepsilon_{si}}{\varepsilon_{ox}} \cdot \frac{t_{ox}}{D} \tag{4.62}$$

where $t_{ox}$ is the gate oxide thickness, $D$ is the channel depletion width, $\varepsilon_{si}$ is the relative permeability of silicon, and $\varepsilon_{ox}$ is the relative permeability of oxide.

Switching power $P_{sw}$ is a consequence of the gate input signal transition causing a charge or discharge of certain internal capacitances. A simple estimate is

$$P_{sw} = I_{sw} \cdot V_d \qquad (4.63)$$

where $I_{sw}$ is the average switching current flow, which can be obtained through circuit simulation [99].

## 4.6 CELL RELIABILITY ESTIMATION

SPICE simulation can provide continuous output of these stress parameters such as voltage and current. For reliability estimation involving multiple failure mechanisms, the simple average stress over time is not accurate enough because most of the failure models have exponential or power law dependence on voltage or temperature. By choosing an appropriate sampling, interval $T_I$, the continuous stress profile can be discretized by sampling periodically. Assuming each operation takes one clock cycle $T_A$, for transistor $M_m$ ($m = 1, 2, \ldots, M$) in the cell, the stress profile in $j$th period ($j = 1, 2, \ldots, [T_A/T_I]$) is $V_{gs}^{ij}$, $V_{ds}^{ij}$, and $I_{ds}^{ij}$. For interconnect $W_w$ ($w = 1, 2, \ldots, W$), the stress current in $j$th period is $I_w^{kj}$, and the stress temperature is $T^j$.

### 4.6.1 ESF Evaluation

ESFs are specific to cell, transistor, operation, and failure mechanism. In general, cell transistors' degradation is not the same because each transistor may have its unique stress profile. For an example, HCI is normally observed in NMOS transistors while NBTI is only observed in PMOS transistors.

Given a transistor in a cell, with the extracted operation-specific dynamic stress profile, failure mechanism ESF estimation takes the following steps. To simplify the derivation, let's take a general transistor $M_0$ as an example. For $M_0$, voltage $V_{gs}^j$, drain to source voltage $V_{ds}^j$, and the stress temperature $T^j$. A standard stress profile is set as voltage:

$$V_{gs}^S = V_{ds}^S = V_D \qquad (4.64)$$

where $V_D$ is the supply voltage and temperature $T_s$.

1. **TDDB ESF**

   Assume the TDDB lifetime dependence on voltage is modeled by the exponential law with voltage coefficient $\gamma_{\text{TDDB}}$ and dependence on temperature is model by Arrhenius relationship with activation energy $E_{a\text{TDDB}}$. The TDDB ETST can be calculated as

   $$T_{\text{TDDB}}^E = \sum_{j=1}^{J} T_I \cdot \exp\left(\gamma_{\text{TDDB}} \cdot \left(V_g^j - V_D\right)\right) \cdot \exp\left(\frac{E_{a\text{TDDB}}}{K}\left(\frac{1}{T_S} - \frac{1}{T_j}\right)\right)$$

   (4.65)

   and the ESF is

   $$\text{ESF}_{\text{TDDB}} = \frac{T_{\text{TDDB}}^E}{T_A}$$

   (4.66)

2. **HCI ESF**

   For HCI, lifetime dependence on voltage is modeled by the empirical exponential model with coefficient $\gamma_{\text{HCI}}$. The temperature dependence is modeled by the Arrhenius relationship with activation energy $E_{a\text{HCI}}$. The HCI ETST can be calculated as

   $$T_{\text{HCD}}^E = \sum_{j=1}^{J} T_I \cdot \exp\left(\gamma_{\text{HCD}} \cdot \left(\frac{1}{V_D} - \frac{1}{V_d^j}\right)\right) \cdot \exp\left(\frac{E_{a\text{HCD}}}{K}\left(\frac{1}{T_S} - \frac{1}{T_j}\right)\right)$$

   (4.67)

   and HCI ESF is

   $$\text{ESF}_{\text{HCD}} = \frac{T_{\text{HCD}}^E}{T_A}$$

   (4.68)

3. **NBTI ESF**

   NBTI degradation only happens to the PMOS transistor. Assume NBTI voltage dependence is modeled by the exponential model with voltage coefficient $\gamma_{\text{NBTI}}$ and activation energy of temperature acceleration is $E_{a\text{NBTI}}$. ETST can be calculated as

   $$T_{\text{HCD}}^E = \sum_{j=1}^{J} T_I \cdot \exp\left(\gamma_{\text{NBTI}} \cdot \left(V_g^j - V_D\right)\right) \cdot \exp\left(\frac{E_{a\text{NBTI}}}{K}\left(\frac{1}{T_S} - \frac{1}{T_j}\right)\right)$$

   (4.69)

   and the ESF is

   $$\text{ESF}_{\text{NBTI}} = \frac{T_{\text{NBTI}}^E}{T_A}$$

   (4.70)

For cell interconnect, EM ESF can be calculated by the same way. Assume interconnect $W_0$ has a stress current $J_j$ in $T_j$. EM activation energy is $E_{aEM}$, and the standard stress current is $J_S$ from Black's equation. Assuming the current density power is $ii$, the ETST can be calculated as

$$T^E_{EM} = \sum_{j=1}^{J} T_I \cdot \left(\frac{J_j}{J_S}\right)^n \cdot \exp\left(\frac{E_{aEM}}{K}\left(\frac{1}{T_S} - \frac{1}{T_j}\right)\right) \qquad (4.71)$$

and the ESF is

$$\text{ESF}_{EM} = \frac{T^E_{EM}}{T_A} \qquad (4.72)$$

### 4.6.2 Cell Reliability

Cell transistor reliability can be estimated with the transistor operation profile, ESFs, and failure mechanism lifetime distribution. For each failure mechanism, the best-fit lifetime distribution can be decided from acceleration tests.

Assume cell has an $N$ type of operation, and denote $f_i$ as the frequency of the $i$th ($i = 1, \ldots, N$) operation. $F_0$ is the device operating frequency. Transistor $M_0$'s ESFs of the $i$th operation are $\text{ESF}^i_{TDDB}$, $\text{ESF}^i_{HCI}$, and $\text{ESF}^i_{NBTI}$ for TDDB, HCI, and NBTI, respectively. $M_0$'s reliability estimation takes the following steps.

1. **TDDB**

   Weibull distribution is used to model TDDB failures. Denote $\beta$ as the Weibull distribution shape parameter and $\alpha_0$ as the scale parameter under standard stress ($V_D$, $T_s$). $M_0$'s TDDB reliability can be estimated by

$$R^{TDDB}_{M_0}(t) = \exp\left[-\left(\frac{\left(\sum_{i=1}^{N} \text{ESF}^i_{TDDB} \cdot f_i/F_0\right)t}{\alpha_0}\right)^{\beta}\right] \qquad (4.73)$$

2. **HCI**

   Lognormal distribution is normally used to model HCI failures. Denote $\mu_{HCI}$ as the mean and $\sigma_{HCI}$ as lifetime distribution under standard stress, respectively. $M_0$'s HCI reliability can be estimated by

$$R^{HCD}_{M_0}(t) = 1 - \Phi\left(\frac{\ln\left(\left(\sum_{i=1}^{N} \text{ESF}^i_{HCI} \cdot f_i/F_0\right)t\right) - \mu_{HCD}}{\alpha_{HCD}}\right) \qquad (4.74)$$

## 3. NBTI

NBTI failures are often modeled by lognormal distribution. $M_0$'s NBTI reliability can be estimated by

$$R_{M_0}^{\text{NBTI}}(t) = 1 - \Phi\left(\frac{\ln\left(\left(\sum_{i=1}^{N} \text{ESF}_{\text{NBTI}}^i \cdot f_i/F_0\right)t\right) - \mu_{\text{NBTI}}}{\alpha_{\text{NBTI}}}\right) \quad (4.75)$$

where $\mu_{\text{NBTI}}$ is the mean and $\sigma_{\text{NBTI}}$ is the standard deviation of the lognormal distribution. If $M_0$ is an NMOS transistor, its reliability is

$$R_{M_0}(t) = R_{M_0}^{\text{TDDB}}(t) \cdot R_{M_0}^{\text{HCD}}(t) \quad (4.76)$$

If $M_0$ is a PMOS transistor and there is only NBTI degradation:

$$R_{M_0}(t) = R_{M_0}^{\text{TDDB}}(t) \cdot R_{M_0}^{\text{NBTI}}(t) \quad (4.77)$$

Special attention must be paid for PMOS if NBTI and HCI coexist. Both NBTI and HCI cause threshold voltage degradation and the accumulation effect should be considered. With the ESFs, the mean of total threshold voltage degradation can be estimated by

$$\Delta V_{\text{th}}(t) = A_{\text{HCD}} \cdot \left(\left(\sum_{i=1}^{N} \text{ESF}_{\text{HCD}}^i \cdot \frac{f_i}{F_0}\right) \cdot t\right)^{n_{\text{HCD}}} \\ + A_{\text{NBTI}} \cdot \left(\left(\sum_{i=1}^{N} \text{ESF}_{\text{NBTI}}^i \cdot \frac{f_i}{F_0}\right) \cdot t\right)^{n_{\text{NBTI}}} \quad (4.78)$$

where $A_{\text{HCI}}$ and $\sigma_{\text{NBTI}}$ are prefactors of HCI and NBTI under the given standard conditions, respectively. $n_{\text{HCI}}$ and $n_{\text{NBTI}}$ are the power coefficients of HCI and NBTI, respectively. The standard variation $\sigma_{\text{V}}(t)$ is

$$\sigma_{\text{V}}^2(t) = \sigma_{\text{HCD}}^2 \cdot \left(\left(\sum_{i=1}^{N} \text{ESF}_{\text{HCD}}^i \cdot \frac{f_i}{F_0}\right) \cdot t\right)^{2n_{\text{HCD}}} \\ + \sigma_{\text{NBTI}}^2 \cdot \left(\left(\sum_{i=1}^{N} \text{ESF}_{\text{NBTI}}^i \cdot \frac{f_i}{F_0}\right) \cdot t\right)^{2n_{\text{NBTI}}} \quad (4.79)$$

where $\sigma_{HCI}$ and $\sigma_{NBTI}$ are the standard deviation of the prefactors of HCI and NBTI, respectively.

PMOS reliability due to NBTI and HCI is

$$R_{M_0}^{NBTI+HCI}(t) = \Phi\left(\frac{V_{criterion} - \Delta V_{th}(t)}{\sigma V(t)}\right) \tag{4.80}$$

where $V_{criterion}$ is the $V_{th}$ degradation failure criterion.

Cell interconnect reliability: Reliability of cell interconnect $W_0$ can be estimated by the same way. Lognormal distribution is applied with $\mu_{EM}$ as the mean and $\sigma_{EM}$ as the standard deviation. $ESF_{EM}^i$ is the ESF of the $i$th operation.

$$R_{W_0}^{EM}(t) = \Phi\left(\frac{\ln\left(\left(\sum_{i=1}^{N} ESF_{EM}^i \cdot f_i/F_0\right)\cdot t\right) - \mu_{EM}}{\sigma_{EM}}\right) \tag{4.81}$$

Finally, cell reliability can be expressed as

$$R_{cell}(t) = \prod_{m=1}^{M} R_{M_m}(t)\cdot \prod_{w=1}^{W} R_{W_w}(t) \tag{4.82}$$

## 4.7 CHIP RELIABILITY PREDICTION

Device reliability can be expressed as

$$R_{device}(t) = R_{power}(t)\cdot \prod R_{block}(t) \tag{4.83}$$

where $R_{power}(t)$ is power network reliability and $\prod R_{block}(t)$ is functional reliability.

### 4.7.1 Functional Block Reliability

For each functional block, reliability can be estimated by considering the block as a series system with cell as the component:

$$R_{block}(t) = \prod R_{cell}(t) \tag{4.84}$$

### 4.7.2 Power Network EM Estimation

To estimate power network EM, current waveform and interconnect temperature should be found first. Current waveform can be obtained through SPICE simulation. For power network interconnect $W_p$ ($p = 1, \ldots, P$), stress is divided into $L$ periods, and current density in the $l$th ($l = 1, \ldots, L$) stress period is $J_p^l$. Local interconnect temperature can be calculated by

$$T_p^l = T_S + \frac{J_p^{l^2} \rho_0}{(\mathrm{K}_i/\Theta_{ti})[1 + 0.88t_i/W] - J_p^{l^2}\rho_0\beta_m} \tag{4.85}$$

and the ETST under standard stress conditions is

$$T_{\mathrm{EM}}^p = \sum_{l=1}^{L} T_l \cdot \left(\frac{J_p^l}{J_S}\right)^n \cdot \exp\left(\frac{E_{a\mathrm{EM}}}{\mathrm{K}}\left(\frac{1}{T_S} - \frac{1}{T_p^l}\right)\right) \tag{4.86}$$

$W_p$'s EM reliability can be estimated as

$$R_{W_p}^{\mathrm{EM}}(t) = \Phi\left(\frac{\ln(T_{\mathrm{EM}}^p) - \mu_{\mathrm{EM}}}{\sigma_{\mathrm{EM}}}\right) \tag{4.87}$$

Reliability of the power network is calculated by

$$R_{\mathrm{power}}(t) = \prod_{p=1}^{P} R_{W_p}^{\mathrm{EM}}(t) \tag{4.88}$$

Since we estimate each mechanism to contribute linearly to the total reliability, we can add up the relative proportions of each mechanism in the matrix, where we only need to find the correct proportionality of each based on the multiple HTOL (M-HTOL) testing, as described in Chapter 1.

## 4.8 MATRIX METHOD

A demonstration will show how the matrix method can be used to predict the reliability of a generic 1 billion transistor chip made on 40 and in 28 nm technologies based on the M-HTOL testing. We build a matrix of the ESFs based on the published models of the failure mechanisms. Each time to fail is the average time to a certain predetermined degradation. In this example, 10% was chosen as the extrapolated degradation percent (in ring oscillator degradation). In this example, 1.1 V was determined to be the nominal operating voltage, so acceleration was performed at 1.4 and 1.6 V while the temperatures tested were at 30°C and 130°C, to be sure we had the full range of excitations that would lead to failure.

The relative values are based on a nominal voltage of 1.1 V and a base temperature of 30°C. From that point in the matrix, each AF for each mechanism is normalized as seen in the matrix given in Table 4.6.

We solved this matrix uniquely for the relative weighting of each mechanism by solving the inverse matrix equal to the right-hand column (measured) values as given in Table 4.7.

These results allow us to find the relative contribution for each mechanism based on our M-HTOL test using the multiple matrix solution. From this result, it is very simple to predict the total device reliability based on the contribution from each mechanism given the voltage, temperature, and frequency. Frequency is calculated by multiplying the EM contribution and the HCI contribution by the actual operation frequency divided by the measured ring frequency of 1 GHz. Hence, we can make predictions for the FIT, measured in failures per billion part hours of operation. This number is normalized to the number of gates in the digital chip, as we plot FIT per billion gates per chip. These results are seen in Figures 4.6 and 4.7

**Table 4.6 Relative Accelerations for Four Mechanisms Plus the Measured AF from the Extrapolated Testing**

| Temperature | Volts | TDDB | HCI | BTI | EM | Measured |
|---|---|---|---|---|---|---|
| 130 | 1.1 | 248 | 6 | 4714 | 5184 | 2500 |
| 30 | 1.4 | 7503 | 1339 | 469 | 1 | 2000 |
| 130 | 1.4 | 1,858,326 | 7411 | 2,209,687 | 5184 | 1,450,000 |
| 30 | 1.1 | 1 | 1 | 1 | 1 | 1 |

**Table 4.7 Inverse Matrix from the 40 nm AFs Above with the Resulting Percentages for Each Mechanism on the Right**

| | | Inverse Matrix | | Results |
|---|---|---|---|---|
| 4.35347E − 05 | 0.000168494 | − 2.64664E − 08 | 0.225730514 | 18% TDDB |
| − 0.000231293 | 0.000147774 | − 1.79615E − 08 | 1.199345462 | 30% HCI |
| − 3.63624E − 05 | 0.000141489 | 4.75883E − 07 | 0.186189902 | 50% BTI |
| 0.000224121 | 0.000120769 | − 4.31455E − 07 | 0.159804849 | 2% EM |

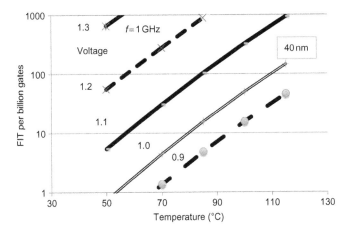

*Figure 4.6 FIT per billion gates at 40 nm versus temperature for different voltages.*

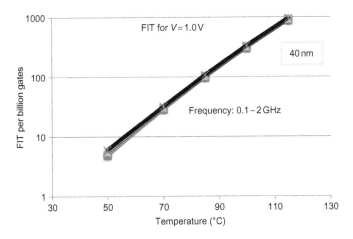

*Figure 4.7 FIT per billion gates at 40 nm versus temperature for different frequencies.*

where we plot the FIT versus temperature for voltages and frequencies in this 40 nm technology.

The most notable result here is that there is nearly no frequency dependence on degradation whereas the temperature and voltage effects are quite dramatic. It is clear that in the 40 nm node, keeping the device cool and operating at a lower voltage would improve the reliability if the performance is not too much degraded.

This same approach was followed with the 28 nm example given in Chapter 1. Figures 4.8 and 4.9 show the similar plots of FIT versus

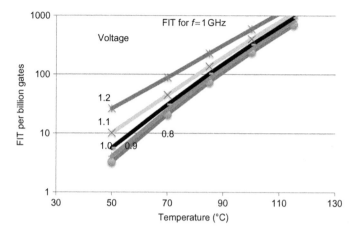

*Figure 4.8 FIT versus temperature for 28 nm devices at 1 GHz operation.*

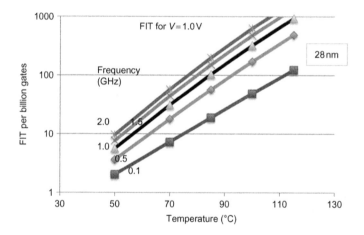

*Figure 4.9 FIT versus temperature for 5 frequencies at 1.0 V nominal voltage.*

temperature for both voltage and frequency. In these cases, since the EM effect is much more dramatic, even nearly 50% of the contribution, you should notice a much higher effect of frequency on reliability.

[1] Bernstein JB, Salemi S, Yang L, Dai J, Qin J. Physics-of-failure based handbook of microelectronic systems. Utica, NY: Reliability Information Analysis Center; 2008. ISBN-10: 1-933904-29-1.

[2] Bernstein JB, Gurfinkel M, Li X, Walters J, Shapira Y, Talmor M. Electronic circuit reliability modeling. Microelectron Reliab 2006;46(12):1957−79.

[3] JEDEC publication. JEP122G, Failure Mechanisms and Models for Semiconductor Devices. (Revision of JEP122F, November 2010 or later).

[4] Goel A, Graves RJ. Electronic system reliability: collating prediction models. IEEE Trans Electron Dev 2006;6(2):258−65.

[5] Bechtold L, Revamping Mil-HDBK-217: Were ready for it, available at: <http://vita-technologies.com/articles/revamping-mil-hdbk-217-we237re-ready-it/> Vita-Technologies, April 25, 2008.

[6] Morris SF, Reilly JF. MIL-HDBK-217—A favorite target. In: Proceedings annual reliability and maintainability symposium; 1993. p. 503−9.

[7] Fuqua NB, Electronic Reliability Prediction, RAC publication. vol. 4, no. 2, Alionscience, 2005, available at: <http://src.alionscience.com/pdf/pred.pdf>.

[8] Qin J. A new physics-of-failure based VLSI circuit reliability simulation and prediction methodology [Dissertation]. Faculty of the Graduate School of the University of Maryland; 2007.

[9] REJ27L0001-0100. Semiconductor reliability handbook (Renesas), Revision Date: August 31, 2006 Rev.1.00.

[10] Bisschop J. Reliability methods and standards. Microelectron Reliab 2007;47(9−11):1330−5.

[11] Reliability Testing Semiconductor Company. Toshiba Leading Innovation.

[12] Kim KK, Reliable CMOS VLSI Design Considering Gate Oxide Breakdown, CENICS 2012: The Fifth International Conference on Advances in Circuits, Electronics and Micro-electronics.

[13] Suehle JS. Ultra thin gate oxide reliability: physical models, statistics, and characterization. IEEE Trans Electron Dev 2002;49(6):958−71.

[14] Walter JD, Bernstein JB. Semiconductor device lifetime enhancement by performance reduction.

[15] Cheung KP. Soft breakdown in thin gate oxide—a measurement artifact. In: 41st annual international reliability symposium (IEEE'03), Dallas, TX; 2003.

[16] Yeo YC, et al. MOSFET gate oxide reliability: anode hole injection model and its application. IJHSE 2001;11(3):849−86.

[17] Hu C. A unified gate oxide reliability model. In: IEEE annual international reliability physics symposium (IRPS'99); 1999. p. 47−51.

[18] Abadeer WW, et al. Key measurements of ultra thin gate dielectric reliability and in-line monitoring. IBM J Res Dev 1999;43(3).

[19] Luo H, et al. The different gate oxide degradation mechanism under constant voltage/current stress and ramp voltage stress. IRW final report; 2000.

[20] JEDEC Publication No. 122B.

[21] Nicollian PE, et al. Experimental evidence for voltage driven breakdown models in ultrathin gate oxide. In: 38th annual international reliability physics symposium (IEEE'00), San Jose, CA; 2000.

[22] Wu EY, et al. Power-law voltage acceleration: a key element for ultra-thin gate oxide reliability. Microelectron Reliab 2005;45:1809−34.

[23] Pompl T, et al. Voltage acceleration of time dependent breakdown of ultra-thin gate dielectrics. Microelectron Reliab 2005;45:1835−41.

[24] Wu EY, et al. CMOS scaling beyond 100-nm node with silicon-dioxide-based gate dielectric. IBM J Res Dev 2002;46(2/3).

[25] Wu EY, et al. Interplay of voltage and temperature acceleration of oxide breakdown for ultra-thin gate oxides. Solid-State Electron 2002;46:1787−98.

[26] Li X, Qin J, Huang B, Zhang X, Bernstein JB. SRAM circuit-failure modeling and reliability simulation with SPICE, Device and Materials Reliability. IEEE Trans June 2006; 6(2):235−46.

[27] Costa UMS, et al. An improved description of the dielectric breakdown in oxides based on a generalized Weibull distribution. Physica A 2006;361:209−15.

[28] Gall M, Capasso C, Jawarani D, Hernandez R, Kawasaki H. Statistical analysis of early failures in electromigration. J Appl Phys 2001;90:732−40.

[29] Sune J, Placencia I, Barniol N, Farres E, Martin F, Aymerich X. On the break down statistics of very thin SiO$_2$ films. Thin Solid Films 1990;185:347−62.

[30] Dumin J, Mopuri SK, Vanchinathan S, Scott RS, Subramoniam R, Lewis TG. High field related thin oxide wearout and breakdown. IEEE Trans Electron Dev 1995;42:760−72.

[31] Mahapatara S, Parikh CD, Rao VR, Viswanathan CR, Vasi J. Device scaling effects on hot-carrier induced interface and oxide-trapping charge distributions in MOSFETs. IEEE Trans Electron Dev 2000;47:789−96.

[32] Groeseneken G, Degraeve R, Nigam T, den bosch GV, Maes H. Hot carrier degradation and time dependent dielectric breakdown in oxides. Microelectron Eng 1999;49:27−40.

[33] Maes H, Groeseneken G, Degraeve R, Blauwe JD, den Bosch GV. Assessment of oxide reliability and hot carrier degradation in CMOS technology. Microelectron Eng 1998;40:147−66.

[34] Hu C, Tam SC, Hsu F-C, Ko P-K, Chan T-Y, Terrill KW. Hot-carrier-induced MOSFET degradation—model, monitor, and improvement. IEEE Trans Electron Dev 1985;375−84 ED-32.

[35] Acovic A, Rosa GL, Sun Y-C. A review of hot-carrier degradation mechanisms in MOSFETs. Microelectron Reliab 1996;36:845−69.

[36] JEDEC Publication. Failure mechanisms and models for semiconductor devices; March 2006.

[37] Hyoen-Seag Kim, Method and apparatus for predicting semiconductor device lifetime. US Patent No. 6873932B1; 2005.

[38] Segura J, Hawkins CF. CMOS electronics: how it works, how it fails. IEEE Press/Wily Interscience 2004.

[39] Groeseneken G, Degraeve R, Nigam T, Van den Bosch G, Maes HE. Hot carrier degradation and time-dependent dielectric breakdown in oxides. Microelectron Eng 1999; 49:27−40.

[40] Takeda E, Kume H, Toyabe T, Asai S. Submicrometer MOSFET structure for minimizing hot-carrier generation. IEEE J Solid-State Circ 1982;SC-17(2).

[41] Haggag A, McMahon W, High performance chip reliability from short-time tests. In: 39th annual international reliability physics symposium (IEEE'01), Orlando, FL; 2001.

[42] Rittman D. Nanometer reliability.

[43] Lee YH, et al. Prediction of logic product due to thin-gate oxide breakdown. IRPS 2006;18−28.

[44] Deal BE. The current understanding of charges in the thermally oxidized silicon structure. J Electrochem Soc 1974;121:198c−205c.

[45] Goetzberger A, et al. On the formation of surface states during aging of thermal Si-SiO$_2$ interfaces. J Electrochem Soc 1973;120:90−6.

[46] Jeppson KO, Svensson CM. Negative bias stress of MOS devices at high electric fields and degradation of MNOS devices. J Appl Phys 1977;48:2004−14.

[47] Schroder DK, Babcock JA. Negative bias temperature instability: road to cross in deep submicron semiconductor manufacturing. J Appl Phys 2003;94:1−18.

[48] Alam MA, Mahapatra S. A comprehensive model of PMOS NBTI degradation. Microelectron Reliab 1973;45:71−8.

[49] Huard V, et al. NBTI degradation: from physical mechanisms to modeling. Microelectron Reliab 2006;46:1−23.

[50] Stathis JH, Zafar S. The negative bias temperature instability in MOS device: a review. Microelectron Reliab 2006;46:270−86.

[51] Schroder DK. Negative bias temperature instability: what do we understand? Microelectron Reliab 2007;47:841−52.

[52] Goetzberger A, et al. On the formation of surface states during aging of thermal Si-SiO$_2$ interfaces. J Electrochem Soc 1973;120:90−6.

[53] Zafar S, et al. Evaluation of NBTI in HfO$_2$ gate-dielectric stacks with tungsten gates. IEEE Electron Device Lett 2004;25(3):153−5.

[54] Zafar S. Statistical mechanics based model for negative bias temperature instability induced degradation. J Appl Phys 2005;97:1−9.

[55] Haller G, Knoll M, Braunig D, Wulf F, Fahrner WR. Bias temperature stress on metal-oxide-semiconductor structures as compared to ionizing irradiation and tunnel injection. J Appl Phys 1984;56:184.

[56] Chaparala P, Shibley J, Lim P. Threshold voltage drift in PMOSFETS due to NBTI and HCI. In: IRW, IEEE; 2000. p. 95−7.

[57] Mahapatra S, et al. Investigation and modeling of interface and bulk trap generation during negative bias temperature instability of p-MOSFETs. IEEE Trans Electron Dev 2004;51 (9):1371−9.

[58] Jensen F. Electronic component reliability. John Wiley & Sons; 1995.

[59] Young D, Christou A. Failure mechanism models for electromigration. IEEE Trans Reliab 1994;43:186−92.

[60] Hau-Riege CS, Thompson CV. Electromigration in Cu interconnects with very different grain structures. Appl Phys Lett 2001;78:3451−543.

[61] Fischer AH, Abel A, Lepper M, Zitzelsbergr A, von Glasow A. Modeling bimodal electromigration failure distributions. Microelectron Reliab 2001;41:445−53.

[62] Yokogawa S, Okada N, Kakuhara Y, Takizawa H. Electromigration performance of multi-level damascence copper interconnects. Microelectron Reliab 2001;41:1409–16.

[63] Gall M, Capasso C, Jawarani D, Hernandez R, Kawasaki H. Statistical analysis of early failures in electromigration. J Appl Phys 2001;90:732–40.

[64] Ohring M. Reliability and failure of electronic materials and devices. Academic Press; 1998.

[65] Bernstein JB, et al. Electronic Circuit Reliability Modeling. Microelectron Reliab 2006;46:1957–79.

[66] Williams MMR, Throne MC. The estimation of failure rates for low probability events. Prog Nucl Energ 1997;31(4):373–476.

[67] Coolen FPA, Coolen-Schrijner P. On zero-failure testing for Bayesian high-reliability demonstration. In: Proc. IMechE, Part O: J Risk Reliab, vol. 220. p. 35–44.

[68] Coolen FPA. On probabilistic safety assessment in the case of zero failures. In: Proc. IMechE, PartO: J Risk Reliab, vol. 220. p. 105–114.

[69] Tobias PA, Trindade DC. Applied reliability. Van Nostrand Reinhold Company; 1986.

[70] Meade DJ. Failure rate estimation in the case of zero failures. In: SPIE, vol. 3216.

[71] Calculating MTTF when you have zero failures. Technical Brief from Relex Software Corporation.

[72] <http://quality.zarlink.com/quality_data/failrate.htm>.

[73] Caruso H, Dasgupta A. A fundamental overview of acceleration-testing analytic model. In: Reliability and maintainability symposium, proceedings, Annual, January 19–22, 1998; p. 389–93.

[74] JEDEC Standard, JESD85. Methods for calculating failure rates in units of FITs. JEDEC Solid State Technology Association; 2001.

[75] White M. Product reliability and qualification challenges with CMOS scaling. AVSI Consortium; 2005.

[76] Society for Quality. Theor Methods, 1995;90(432).

[77] Gurland J, Sethuraman J. Reversal of increasing failure rates when pooling failure data. Technometrics 1994;36 [American Statistical Association and the American Society for Quality Control].

[78] Proschan F. Theoretical explanation of observed decreasing failure rate. Technometrics 1963;42(1) [American Statistical Association and the American Society for Quality].

[79] Block H, Joe H. Tail behavior of the failure rate functions of mixtures. Lifetime Data Analysis 1997;3:269–88 [Kluwer Academic Publishers, Boston. Manufactured in the Netherlands].

[80] Balakrishnan N, Asit PBasu. The exponential distribution. Theory, methods and applications. Gordon and Breach Publishers; 1995.

[81] Murphy KE, Carter CM, Brown S. The exponential distribution: the good, the bad and the ugly. A practical guide to its implementation. In: Proceedings Annual Reliability and Maintainability Symposium; 2002.

[82] Drenick RF. The failure law of complex equipment. J Soc Ind Appl Math 1960;8(4).

[83] Denson W. The history of reliability prediction. IEEE Trans Reliab 1998;47(3-SP):SP321–8.

[84] Foucher B, Boullie J, Meslet B, Das D. A review of reliability prediction methods for electronic devices. Microelectron Reliab 2002;42:1155–62.

[85] MIL-HDBK-217F. Reliability prediction of electronic equipment. Washington, DC: Department of Defense; 1991.

[86] Stathis JH. Reliability limits for the gate insulator in CMOS technology. IBM J Res Dev 2002;46(2/3):265–86.

[87] Haggag A, et al. Reliability projections of product fail shift and statistics due to HCI and NBTI. IRPS 2007; 93-06.

[88] Bornstein W, et al. Field degradation of memory components due to hot carriers. IRPS 2006;294–8.

[89] Rosa GL, et al. IEEE NBTI-channel hot carries effects in PMOSFETs in advanced CMOS technologies. IRPS; 1997.

[90] Zhu QK. Power distribution network design for VLSI. Wiley Interscience; 2004.

[91] Blech IA. Electromigration in thin aluminum films on titanium nitride. J Appl Phys 1976;47:1203–8.

[92] White M, et al. Product reliability trends, derating considerations and failure mechanisms with scaled CMOS. IIRW 2006;156–9.

[93] Rangan S, et al. IEEE Universal recovery behavior of negative bias temperature in stability. IEDM; 2003.

[94] Ershov M, et al. Degradation dynamics, recovery, and characterization of negative bias temperature instability. Microelectron Reliab 2005;45:99–105.

[95] Gunther SH, et al. Managing the impact of increasing microprocessor power consumption. ITJ 2001;Q1:1–9.

[96] Cheng K, et al. Electrothermal analysis of VLSI systems. Kluwer Academic Publishers; 2000.

[97] Skardron K, et al., Temperature-aware microarchitecture. In: 30th International Symposium on Computer Architecture; 2003. p. 2–13.

[98] Bakoglu H. Circuits interconnections, and packaging for VLSI. Addison Wesley Company; 1990.

[99] Uyemura JP. Introduction to VLSI circuits and systems. John Wiley & Sons, Inc; 2002.

Printed and bound by CPI Group (UK) Ltd, Croydon, CR0 4YY

03/10/2024

01040421-0008